International GCSE Chemistry

Set A Paper 1

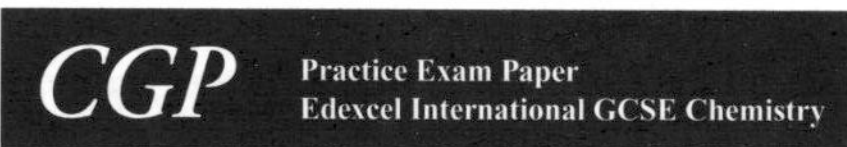

In addition to this paper you should have:
- A ruler.
- A calculator.

Centre name				
Centre number				
Candidate number				

Surname	
Other names	
Candidate signature	

Time allowed:
- 2 hours

Instructions to candidates
- Write your name and other details in the spaces provided above.
- Use a pen with black ink.
- You are allowed to use a calculator.
- Answer **all** questions in the spaces provided.
  There might be more space than you need.
- Answer multiple choice questions by putting a cross in the correct box.
  If you need to change your answer, draw a horizontal line through the box.
  Then mark your new answer as normal.

Information for candidates
- The marks available are given in brackets at the end of each question.
- There are 110 marks available for this paper.

Advice to candidates
- Try to answer all the questions.
- Carefully read each question before you try to answer it.
- If you have time at the end of the exam, check your answers.

For examiner's use

Q	Attempt Nº 1	2	3	Q	Attempt Nº 1	2	3
1				6			
2				7			
3				8			
4				9			
5				10			
				Total			

1

The Periodic Table

Key:

Relative atomic mass		
1		
H		
Hydrogen		
1		

Relative atomic mass → (top number)

Atomic (proton) number → (bottom number)

Group 1	Group 2											Group 3	Group 4	Group 5	Group 6	Group 7	Group 0
																	4 **He** Helium 2
7 **Li** Lithium 3	9 **Be** Beryllium 4											11 **B** Boron 5	12 **C** Carbon 6	14 **N** Nitrogen 7	16 **O** Oxygen 8	19 **F** Fluorine 9	20 **Ne** Neon 10
23 **Na** Sodium 11	24 **Mg** Magnesium 12											27 **Al** Aluminium 13	28 **Si** Silicon 14	31 **P** Phosphorus 15	32 **S** Sulfur 16	35.5 **Cl** Chlorine 17	40 **Ar** Argon 18
39 **K** Potassium 19	40 **Ca** Calcium 20	45 **Sc** Scandium 21	48 **Ti** Titanium 22	51 **V** Vanadium 23	52 **Cr** Chromium 24	55 **Mn** Manganese 25	56 **Fe** Iron 26	59 **Co** Cobalt 27	59 **Ni** Nickel 28	63.5 **Cu** Copper 29	65 **Zn** Zinc 30	70 **Ga** Gallium 31	73 **Ge** Germanium 32	75 **As** Arsenic 33	79 **Se** Selenium 34	80 **Br** Bromine 35	84 **Kr** Krypton 36
85 **Rb** Rubidium 37	88 **Sr** Strontium 38	89 **Y** Yttrium 39	91 **Zr** Zirconium 40	93 **Nb** Niobium 41	96 **Mo** Molybdenum 42	98 **Tc** Technetium 43	101 **Ru** Ruthenium 44	103 **Rh** Rhodium 45	106 **Pd** Palladium 46	108 **Ag** Silver 47	112 **Cd** Cadmium 48	115 **In** Indium 49	119 **Sn** Tin 50	122 **Sb** Antimony 51	128 **Te** Tellurium 52	127 **I** Iodine 53	131 **Xe** Xenon 54
133 **Cs** Caesium 55	137 **Ba** Barium 56	139 **La** Lanthanum 57	178 **Hf** Hafnium 72	181 **Ta** Tantalum 73	184 **W** Tungsten 74	186 **Re** Rhenium 75	190 **Os** Osmium 76	192 **Ir** Iridium 77	195 **Pt** Platinum 78	197 **Au** Gold 79	201 **Hg** Mercury 80	204 **Tl** Thallium 81	207 **Pb** Lead 82	209 **Bi** Bismuth 83	209 **Po** Polonium 84	210 **At** Astatine 85	222 **Rn** Radon 86
223 **Fr** Francium 87	226 **Ra** Radium 88	227 **Ac** Actinium 89	261 **Rf** Rutherfordium 104	262 **Db** Dubnium 105	266 **Sg** Seaborgium 106	264 **Bh** Bohrium 107	277 **Hs** Hassium 108	268 **Mt** Meitnerium 109	271 **Ds** Darmstadtium 110	272 **Rg** Roentgenium 111							

The Lanthanides (atomic numbers 58-71) and the Actinides (atomic numbers 90-103) are not shown in this table.

Answer **all** questions in the spaces provided

1   This question is about elements and the periodic table.
The periodic table below has eight elements labelled.

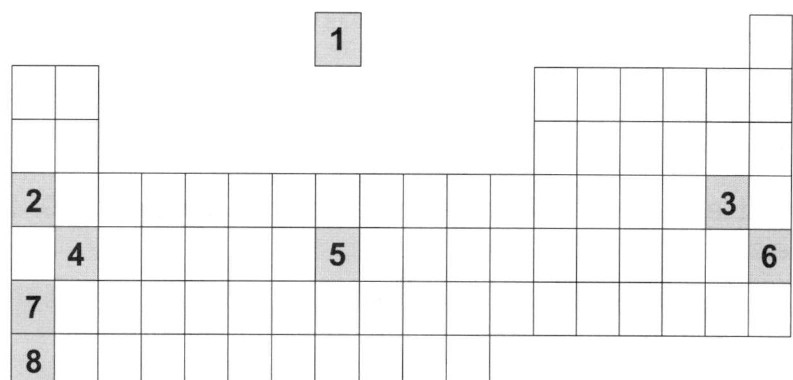

(a) Which of the eight labelled elements have the same number of electrons in their outer shells?

    ☐ **A**   2 and 3

    ☐ **B**   2, 7 and 8

    ☐ **C**   4, 5 and 6

    ☐ **D**   1 and 6

(1)

(b) Which of the eight labelled elements has eight electrons in its outer shell?

    ☐ **A**   1

    ☐ **B**   4

    ☐ **C**   6

    ☐ **D**   8

(1)

(c) What would be the charge on an ion of the element labelled **3** in the table?

    ☐ **A**   −1

    ☐ **B**   −2

    ☐ **C**   +1

    ☐ **D**   +2

(1)

(d) Which of the eight labelled elements would be least likely to take part in a chemical reaction?

☐ **A** 1

☐ **B** 2

☐ **C** 6

☐ **D** 8

(1)

(e) The element labelled **2** in the table has an atomic number of 19.
State its electronic configuration.

....

(1)

(f) Give the numbers of three of the labelled elements that you would expect to have basic oxides.

....

(1)

(Total for Question 1 = 6 marks)

Leave
blank

2 A student prepared a solution of zinc nitrate ($Zn(NO_3)_2$).

(a) This is the method used.

1. Gently warm 20 cm^3 of dilute nitric acid using a Bunsen burner, then turn off the Bunsen burner.

2. Add an excess of zinc oxide to the acid a bit at a time.

3. Filter out the excess solid to get a solution of zinc nitrate.

(i) Complete and balance the equation for the reaction between nitric acid and zinc oxide.

$$ZnO + \text{................................} \rightarrow Zn(NO_3)_2 + \text{..................................}$$

(2)

(ii) What name is given to this type of reaction that forms a salt?

....

(1)

(iii) Explain how the student could tell when the reaction had finished.

....

....

(1)

(iv) Describe how the student could obtain a pure, dry sample of zinc nitrate from the zinc nitrate solution.

....

....

....

....

....

....

(4)

(b) The student used the same method to prepare copper chloride from copper oxide and dilute hydrochloric acid. The equation for this reaction is shown.

$$CuO + 2HCl \rightarrow CuCl_2 + H_2O$$

(i) Calculate the maximum mass of copper chloride that could be produced if the student used 10.1 g of copper oxide and excess hydrochloric acid.

Maximum mass = ... g

(3)

(ii) The student did the reaction again with a different mass of copper oxide. He calculated that the reaction should produce 12 g of copper chloride. The actual mass of copper chloride produced was 8.4 g.

Calculate the percentage yield of this second reaction.

Percentage yield = ...%

(1)

(Total for Question 2 = 12 marks)

3   Carbon can exist in several different forms.

(a) C_{60} fullerene is one of the naturally occurring forms of carbon.

(i)   Describe the structure and bonding in a C_{60} fullerene molecule.

....

....

....

(2)

(ii) C_{60} fullerene is a simple molecular substance but it
is solid at room temperature. Explain why this is.

....

....

....

....

(3)

(b)   Graphite is another form of carbon, which is a soft, crumbly solid at
room temperature.

Explain why graphite is soft.

....

....

....

....

(2)

(c) Almost all of the carbon atoms found in nature are either ^{12}C or ^{13}C.

Leave blank

(i) Give the term that is used to describe different atomic forms of the same element, which have the same number of protons but a different number of neutrons.

....

(1)

(ii) ^{12}C has a relative abundance of 99%.
The relative atomic mass of carbon is 12.

Show how the relative atomic mass is calculated.

Your answer should clearly show your working.

....

....

....

(2)

(Total for Question 3 = 10 marks)

4 Decane is a long hydrocarbon.

The figure shows a molecule of decane.

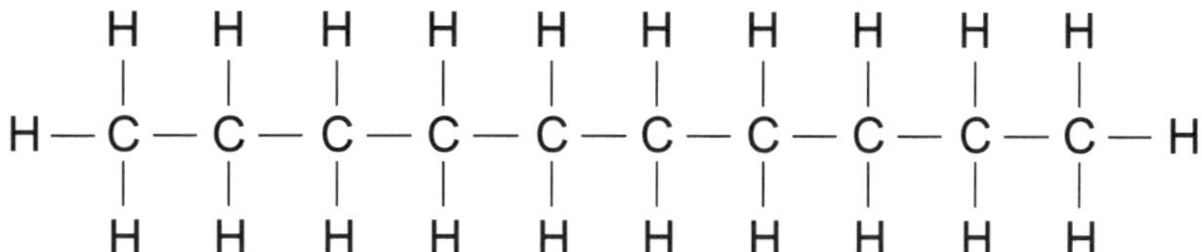

(a) Name the homologous series that decane belongs to.

....

(1)

(b) Decane can be broken down by cracking.

Complete the equation showing one reaction that can take place during the cracking of decane.

$$C_{10}H_{22} \rightarrow C_7H_{16} + \text{.................................}$$

(1)

(c) Propane can also be made by cracking decane.
Propane is a short hydrocarbon.

Describe how propane is produced from the catalytic cracking of decane. Include any necessary reaction conditions.

....

....

....

....

....

....

(3)

(d) Describe a test that could be used to distinguish between the two types of molecule produced by cracking.

You should include the observations that you would expect to see.

....

....

....

....

....

(3)

(e) The graph shows the supply and demand of various crude oil fractions.

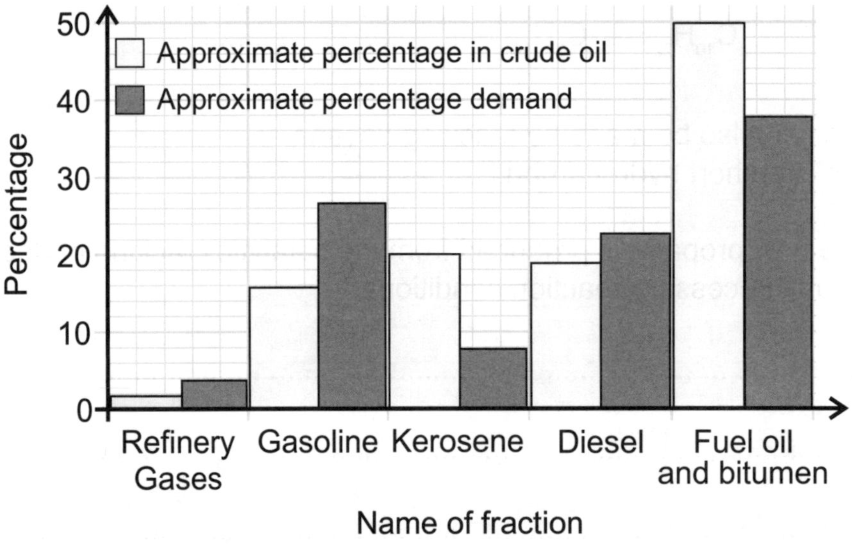

Approximate percentage in crude oil
Approximate percentage demand
Percentage
50
40
30
20
10
0
Refinery Gases
Gasoline
Kerosene
Diesel
Fuel oil and bitumen
Name of fraction

Decane is found is the kerosene fraction of crude oil.

Explain why molecules in the kerosene fraction are cracked.
Use the information from the graph to support your answer.

....

....

....

....

....

(3)

(Total for Question 4 = 11 marks)

5 A student investigated how concentration affects reaction rate by measuring the volume of gas produced by reactions of metals and metal carbonates with acids.

(a) First, the student reacted a small piece of magnesium ribbon with dilute sulfuric acid.

The equation for this reaction is:

$$Mg + H_2SO_4 \rightarrow MgSO_4 + H_2$$

(i) Draw and label an experimental set-up that the student could have used to collect and measure the volume of gas produced over time.

(3)

(ii) During the reaction, the student collected 150 cm³ of hydrogen gas.

150 cm³ of hydrogen gas contains 0.0063 moles of hydrogen.

Suggest why the student may have decided not to measure the rate of reaction by measuring the mass of gas lost from the reaction.

....

....

....

(2)

(b) The student then reacted calcium carbonate with an excess of hydrochloric acid.
The equation for this reaction is:

$$CaCO_3 \ + \ 2HCl \ \rightarrow \ CaCl_2 \ + \ H_2O \ + \ CO_2$$

The student plotted the results of this experiment on the graph below.

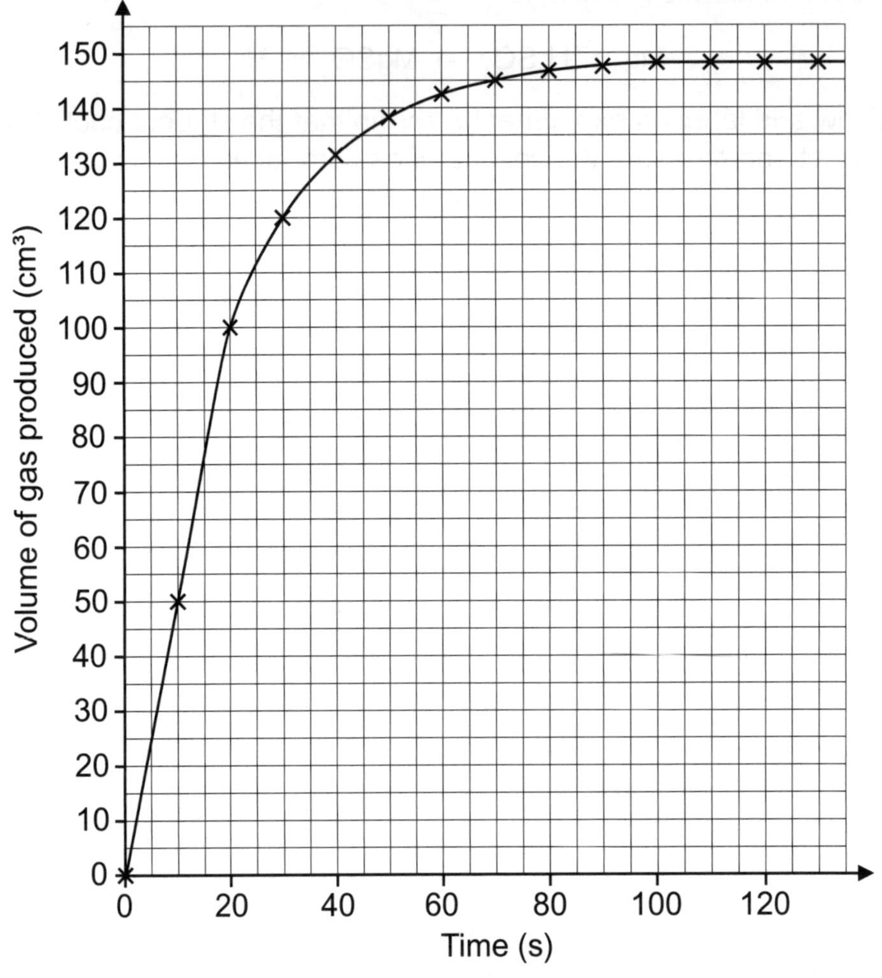

(i) Calculate the volume of gas (in cm³) that was released by the reaction between 20 and 70 seconds.

volume = ... cm³

(1)

(ii) Use the graph to calculate the rate of reaction at 30 seconds.

Give your answer to two significant figures.

rate = ... cm³/s

(3)

(iii) The reaction was repeated using a lower concentration of hydrochloric acid, with all other variables kept the same.

Use collision theory to explain how this would affect the rate of the reaction.

....

....

....

....

(3)

(iv) Sketch how the results of the repeated experiment should look.

Use the axes on the previous page.

(2)

(Total for Question 5 = 14 marks)

6   Analytical tests can be used to identify ionic compounds.

(a) A student mixed sodium hydroxide with a solution of copper(II) chloride.

(i)   State what you would expect to observe during this reaction.

....

(1)

(ii) Complete and balance the equation for the reaction between copper(II) chloride and sodium hydroxide.

$CuCl_2$ + ... $\rightarrow$ .... + ...

(2)

(b) Describe a test that could be used to identify if water is present in a solution.

Include any observations you would expect to see during this test.

....

....

....

....

(3)

(c) A student carries out a flame test on an unknown ionic compound.

This is the method used:

- Take a platinum wire loop and dip it in the sample.
- Place the loop in the blue flame from a Bunsen burner.
- Record the colour of the flame.

(i) Identify **one** error in the student's method.

Explain **one** problem this error could cause.

....

....

....

(2)

(ii) The student repeats the flame test using a correct method.

He also carries out further tests on samples of the compound.

He makes the following observations:

- When the compound is placed in a Bunsen flame, it gives a red colour.
- If dilute HCl is added to a solution of the compound, followed by barium chloride solution, a white precipitate is produced.

Identify the ionic compound by its chemical name.

....

(2)

(Total for Question 6 = 10 marks)

7   A student is separating the compounds found in leaves from a certain plant using paper chromatography.

The first time she carries out the experiment, she finds that two of the substances haven't separated properly on the chromatogram.

(a) Suggest **one** thing the student could do to improve the separation of the spots.

....

....

(1)

The student repeats the experiment to improve the separation of the spots.

The results of this experiment are shown below.

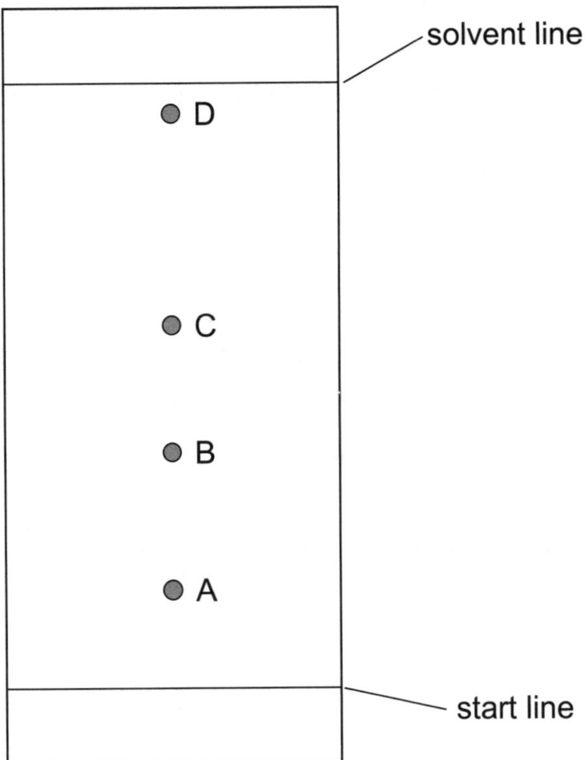

(b) 'Chlorophyll a' is a compound found in leaves. The standard R_f value of chlorophyll a in the solvent that the student used is 0.60.

Identify which spot on the chromatogram represents chlorophyll a.

Show your working.

chlorophyll a = ...

(3)

The student extracts the chlorophyll a from the chromatography paper.

This is the method used:

*   Cut out the portion of paper containing the correct spot.
*   Place the paper in the same solvent as the chromatography experiment was carried out in. The spot of substance will dissolve in the solvent.
*   Evaporate the solution in order to crystallise the substance from the solvent.

(c) Explain how the student could test whether the crystals of the chlorophyll a she has extracted are pure.

....

....

....

(2)

(d) The student repeats the original chromatography experiment using the same solvent.

The solvent travelled 12 cm up the chromatography paper.

Calculate how far **Spot A** would have travelled in this experiment.

Give your answer to 2 significant figures.

distance travelled = ... cm

(3)

(Total for Question 7 = 9 marks)

8   Propene can undergo a variety of different reactions.

$$C_3H_8$$

$+ H_2$
$+$ catalyst

$+$ pressure
$+$ catalyst

$C_3H_6Br_2$ ← $+ Br_2$ — $H_2C=CH\text{-}CH_3$ → Polymer A

propene

(a) The diagram above shows some of the reactions that propene can take part in.

(i)   Name the type of reaction that occurs when Br_2 reacts with propene to form $C_3H_6Br_2$.

....

(1)

(ii) Which property of propene means that it can undergo these types of reaction?

....

(1)

(iii) Draw the displayed formula for C_3H_8.

(1)

(iv) Calculate the M_r of $C_3H_6Br_2$.

molecular mass = ... g/mol

(1)

(v) Draw the repeat unit for **Polymer A**.

(2)

(vi) State **one** problem with disposing polymers by burning.

....

(1)

(b) The diagram shows the displayed formula of another alkene, **B**.

$$H-\overset{\displaystyle H}{\underset{\displaystyle H}{C}}-\overset{\displaystyle H}{C}=\overset{\displaystyle H}{C}-\overset{\displaystyle H}{\underset{\displaystyle H}{C}}-\overset{\displaystyle H}{\underset{\displaystyle H}{C}}-H$$

(i) What is the name of this alkene?

....

(1)

(ii) Write a balanced symbol equation for the complete combustion of alkene **B**.

....

(2)

(Total for Question 8 = 10 marks)

19

9 A student is doing an experiment using iron wool to determine the proportion of oxygen in the air.

Their experimental set-up is shown below.

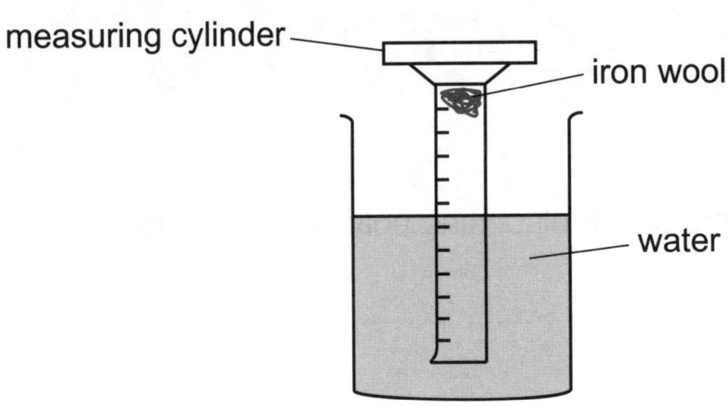

(a) The student measured the volume of air in the measuring cylinder at the start of the experiment. They then measured the volume again every day until they got the same reading for three days running.

(i) Why did the student stop the experiment after three days with the same reading?

....

....

....

(1)

(ii) The volume of air in the measuring cylinder at the start of the experiment was 32.0 cm³. The final volume of air was 25.6 cm³ .

Use these results to calculate the percentage volume of oxygen in the air.

percentage volume of oxygen: ....%

(2)

(iii) The student repeats the experiment, but this time they dip the wool in acetic acid before putting it in the measuring cylinder. The acetic acid catalyses the reaction between the iron and the oxygen in the air.

How would the acetic acid affect both the overall change in volume and the time taken for the volume to stop changing, if at all?

Explain your answer.

....

....

....

....

....

....

(3)

(b) Rust forms when iron reacts with oxygen in the air and water.

(i) Name the compound that is produced when iron rusts.

....

(1)

(ii) How does paint stop iron pipes from rusting?

☐ **A** The layer of paint reacts instead of the iron.

☐ **B** The layer of paint keeps out the reactants that cause iron to rust.

☐ **C** The layer of paint reacts with the iron so it can't rust.

☐ **D** The layer of paint stops flakes of rust from falling off.

(1)

(iii) Give another barrier method that can be used to stop iron rusting.

....

(1)

(iv) Iron can also be galvanised to prevent rusting.

Name the metal that is applied to iron during galvanisation.

....

(1)

(Total for Question 9 = 10 marks)

10 A student wanted to make a chemical pack to use as an alternative to an ice pack for sports injuries.

The pack contains a solid that produces an endothermic reaction when it dissolves in water. A diagram is shown below.

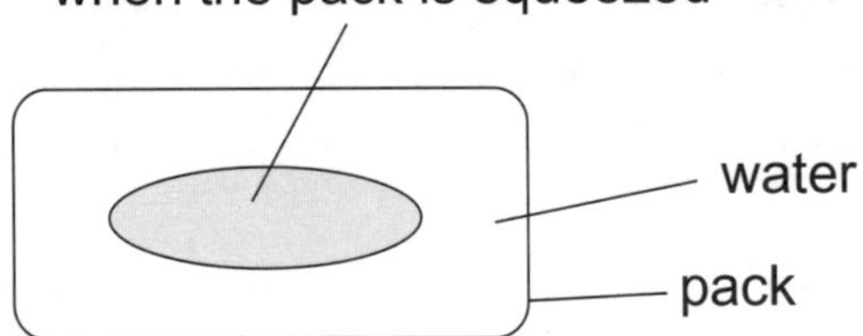

The student mixed the same number of moles of different powdered solids with 50 g of water in a plastic cup. They then measured the temperature change. The results are shown in the table.

Chemical used	Temp. at start (°C)	Temp. at end (°C)
ammonium nitrate	22	7
anhydrous copper sulfate	22	31
potassium nitrate	22	15
sodium chloride	22	19

(a) Evaluate the suitability of the four chemicals the student tested for use in the chemical pack.

....

....

....

....

....

....

....

(4)

(b) The student insulated all four containers with cotton wool.
Explain why it was important that the student insulated the containers.

....

....

....

(2)

(c) Name **two** other things the student did that allowed her to fairly compare
the temperature change in the reactions.

1. ....

2. ....

(2)

(d) Calculate the heat energy change, Q, for the experiment using potassium nitrate.
Give your answer in kJ.

Specific heat capacity of water = 4.2 J/g/°C

Q = ... kJ

(3)

(e) The molar enthalpy change, ΔH, for dissolving sodium chloride is +4.0 kJ/mol.
Calculate the heat energy change that would occur if 25 g of sodium chloride
was dissolved in water.

Q = ... kJ

(3)

(f) The student used 5 g of ammonium nitrate, NH_4NO_3, in the experiment. Calculate the number of moles of ammonium nitrate used.

Amount of ammonium nitrate = ... moles

(2)

(g) State **two** physical properties that you would expect potassium nitrate to have (other than solubility in water).

....

....

(2)

(Total for Question 10 = 18 marks)

TOTAL FOR PAPER = 110 MARKS

International GCSE Chemistry

Set A Paper 2

In addition to this paper you should have:
- A ruler.
- A calculator.

Centre name			
Centre number			
Candidate number			

Surname	
Other names	
Candidate signature	

Time allowed:
- 1 hour 15 minutes

Instructions to candidates
- Write your name and other details in the spaces provided above.
- Use a pen with black ink.
- You are allowed to use a calculator.
- Answer **all** questions in the spaces provided.
  There might be more space than you need.
- Answer multiple choice questions by putting a cross in the correct box.
  If you need to change your answer, draw a horizontal line through the box.
  Then mark your new answer as normal.

Information for candidates
- The marks available are given in brackets at the end
  of each question.
- There are 70 marks available for this paper.

Advice to candidates
- Try to answer all the questions.
- Carefully read each question before you try to answer it.
- If you have time at the end of the exam, check your answers.

For examiner's use

Q	Attempt Nº 1	2	3	Q	Attempt Nº 1	2	3
1				6			
2				7			
3				8			
4				9			
5							
			Total				

The Periodic Table

		Group 1	Group 2											Group 3	Group 4	Group 5	Group 6	Group 7	Group 0
																			4 He Helium 2
		7 Li Lithium 3	9 Be Beryllium 4											11 B Boron 5	12 C Carbon 6	14 N Nitrogen 7	16 O Oxygen 8	19 F Fluorine 9	20 Ne Neon 10
		23 Na Sodium 11	24 Mg Magnesium 12											27 Al Aluminium 13	28 Si Silicon 14	31 P Phosphorus 15	32 S Sulfur 16	35.5 Cl Chlorine 17	40 Ar Argon 18
		39 K Potassium 19	40 Ca Calcium 20	45 Sc Scandium 21	48 Ti Titanium 22	51 V Vanadium 23	52 Cr Chromium 24	55 Mn Manganese 25	56 Fe Iron 26	59 Co Cobalt 27	59 Ni Nickel 28	63.5 Cu Copper 29	65 Zn Zinc 30	70 Ga Gallium 31	73 Ge Germanium 32	75 As Arsenic 33	79 Se Selenium 34	80 Br Bromine 35	84 Kr Krypton 36
		85 Rb Rubidium 37	88 Sr Strontium 38	89 Y Yttrium 39	91 Zr Zirconium 40	93 Nb Niobium 41	96 Mo Molybdenum 42	98 Tc Technetium 43	101 Ru Ruthenium 44	103 Rh Rhodium 45	106 Pd Palladium 46	108 Ag Silver 47	112 Cd Cadmium 48	115 In Indium 49	119 Sn Tin 50	122 Sb Antimony 51	128 Te Tellurium 52	127 I Iodine 53	131 Xe Xenon 54
		133 Cs Caesium 55	137 Ba Barium 56	139 La Lanthanum 57	178 Hf Hafnium 72	181 Ta Tantalum 73	184 W Tungsten 74	186 Re Rhenium 75	190 Os Osmium 76	192 Ir Iridium 77	195 Pt Platinum 78	197 Au Gold 79	201 Hg Mercury 80	204 Tl Thallium 81	207 Pb Lead 82	209 Bi Bismuth 83	209 Po Polonium 84	210 At Astatine 85	222 Rn Radon 86
		223 Fr Francium 87	226 Ra Radium 88	227 Ac Actinium 89	261 Rf Rutherfordium 104	262 Db Dubnium 105	266 Sg Seaborgium 106	264 Bh Bohrium 107	277 Hs Hassium 108	268 Mt Meitnerium 109	271 Ds Darmstadtium 110	272 Rg Roentgenium 111								

Key:

1 H Hydrogen 1

Relative atomic mass →

Atomic (proton) number →

The Lanthanides (atomic numbers 58-71) and the Actinides (atomic numbers 90-103) are not shown in this table.

Edexcel International GCSE Chemistry / Set A / Paper 2

1 Carbon dioxide is a simple molecule whose displayed formula is shown below.

$$O\!=\!C\!=\!O$$

(a) How many covalent bonds are in a molecule of carbon dioxide?

....

(1)

Carbon dioxide is produced when a fuel reacts with oxygen in
a combustion reaction.

(b) Complete the dot and cross diagram for a molecule of oxygen.

   Show the outer electrons of the atoms only.

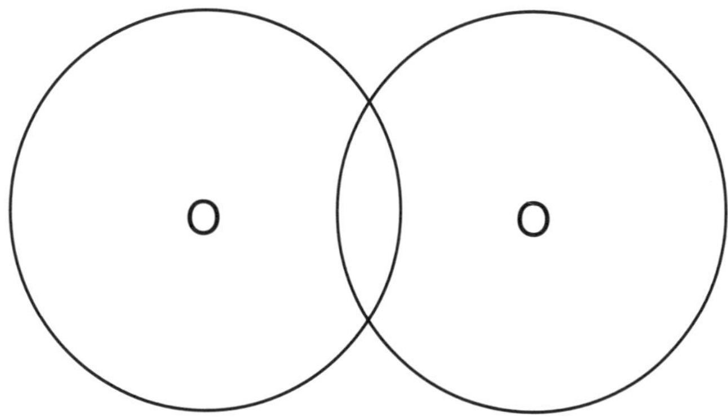

(2)

(c) Some data about the properties of oxygen is shown in the table.

Melting point	Boiling point
−218 °C	−183 °C

Predict the physical state of oxygen at −198 °C.

....

(1)

(d) Predict whether carbon dioxide will have a higher or lower
boiling point than oxygen.

   Explain your answer.

....

....

....

(3)

3

(e) Carbon dioxide can also be produced by heating zinc carbonate ($ZnCO_3$).
The equation for the reaction is:

$$ZnCO_3 \rightarrow ZnO + CO_2$$

A student heated some zinc carbonate and collected the carbon dioxide
given off in a syringe.

The volume of carbon dioxide collected was measured to be 105.6 cm^3 at RTP.

Calculate the mass of zinc carbonate that thermally decomposed.
The molar volume of carbon dioxide at RTP is 24 dm^3.

Moles of $ZnCO_3$ = ... mol

(3)

(Total for Question 1 = 10 marks)

2 The apparatus below can be used in four different methods for separating mixtures.

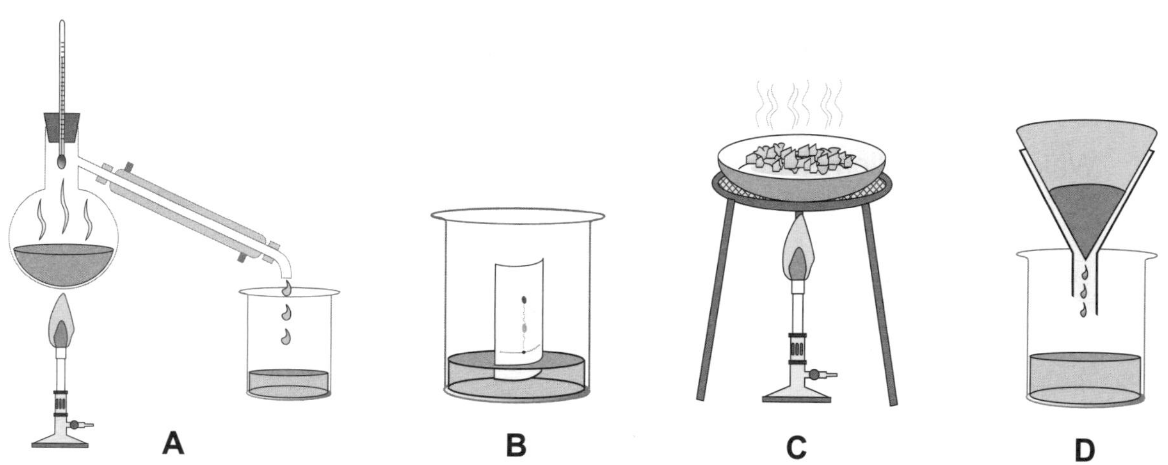

(a) Which of the separation methods would be most suitable for obtaining a pure, dry sample of sodium chloride from a solution of sodium chloride dissolved in water?

☐ A

☐ B

☐ C

☐ D

(1)

(b) Which of the separation methods would be most suitable for separating a mixture of two alcohols with very different boiling points?

☐ A

☐ B

☐ C

☐ D

(1)

(c) Which of the separation methods would be most suitable for separating a mixture of dyes present in an ink?

☐ A

☐ B

☐ C

☐ D

(1)

Leave blank

(d) Rock salt is a mixture of a soluble salt and insoluble sand.

A student wants to obtain a pure, dry sample of the soluble salt from a sample of rock salt in water.

Which combination of the separation methods would be most suitable for the student to use to obtain the soluble salt from the mixture?

   ☐ **A**   D then A

   ☐ **B**   D then C

   ☐ **C**   A, then D, then C

   ☐ **D**   B then A

(1)

(e) Name the separation method that would be most suitable for separating a mixture of liquids that have similar boiling points.

....

(1)

(Total for Question 2 = 5 marks)

3 Silver is a metallic element.

(a) Draw a labelled diagram to show the bonding in silver.

(3)

(b) Explain why silver is a good conductor of electricity.

....

....

(1)

(c) Explain how the structure of silver means that it can be bent and shaped.

....

....

(1)

(Total for Question 3 = 5 marks)

4   When a small piece of sodium is added to water it fizzes around on the surface before dissolving to form a colourless solution.

(a) Write down the word equation for this reaction.

....

(1)

(b) At the end of the reaction, the solution is tested with Universal indicator.

The solution turns purple.

What does this tell you about the solution?

....

(1)

Potassium is below sodium in Group 1 of the periodic table.

(c) Compare the reactivity of sodium and potassium.

Explain your answer.

....

....

....

....

(3)

(Total for Question 4 = 5 marks)

5 An empirical formula shows the ratio of atoms of each element in a compound.

(a) A student carried out an experiment to find the empirical formula of magnesium oxide.

The student set up their experiment as shown in the diagram and then followed the method below.

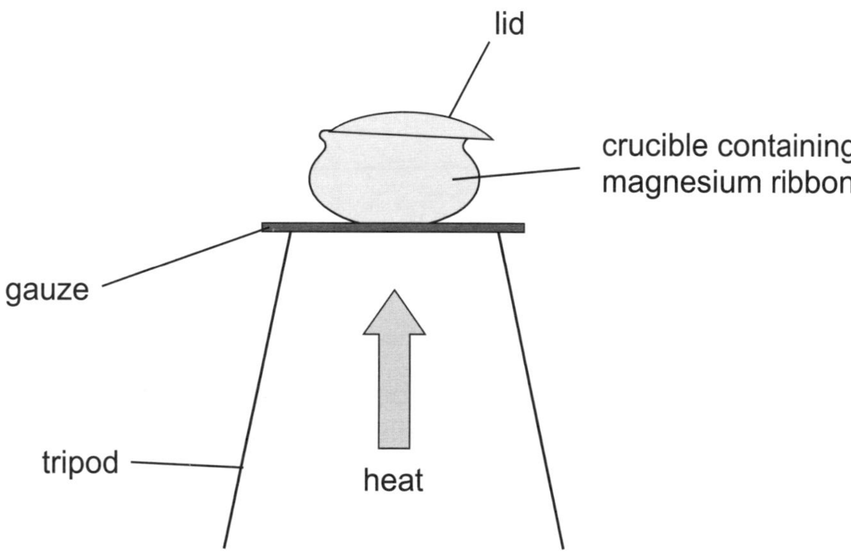

- Heat a crucible until hot to dry it.
- Leave the crucible to cool, then weigh it along with its lid.
- Add a clean magnesium ribbon to the crucible and reweigh.
- Heat the crucible strongly for 10 minutes with the lid on, leaving a small gap to allow oxygen to enter the crucible.
- Allow the crucible and lid to cool then reweigh, recording the final mass.

The results of the student's experiment are shown in the table.

Mass of crucible and lid	35.25 g
Mass of crucible, lid and magnesium ribbon	35.97 g
Mass of crucible, lid and magnesium oxide	36.45 g

(i) Use the data in the table to find the empirical formula of magnesium oxide.

Show your working.

(5)

(ii) A second student does the same experiment, but the final mass of the crucible, lid and magnesium oxide is lower than expected. Suggest why this may have occurred.

....

(1)

(b) You can also find empirical formulae by reducing metal oxides.

Draw an experimental set-up you could use to carry out the reduction of copper oxide, in order to find its empirical formula.

(2)

(Total for Question 5 = 8 marks)

6 A student carried out a titration to work out the concentration of a sample of potassium hydroxide solution.

The apparatus that the student used is shown in the diagram.

This is the method used.

1. Use a measuring cylinder to measure 50.0 cm³ of potassium hydroxide solution.
2. Add the potassium hydroxide solution to a conical flask.
3. Add two or three drops of phenolphthalein solution to the conical flask.
4. Use a funnel to fill a burette with 0.0250 mol/dm³ sulfuric acid.
5. Record the initial volume of the acid in the burette.
6. Add the acid to the alkali a bit at a time. Give the conical flask a regular swirl. Go especially slowly when the end-point is about to be reached.
7. Record the final volume of acid in the burette at the end-point. Use this, along with the initial reading, to calculate the volume of acid used to neutralise the alkali.
8. Repeat the experiment at least two more times.

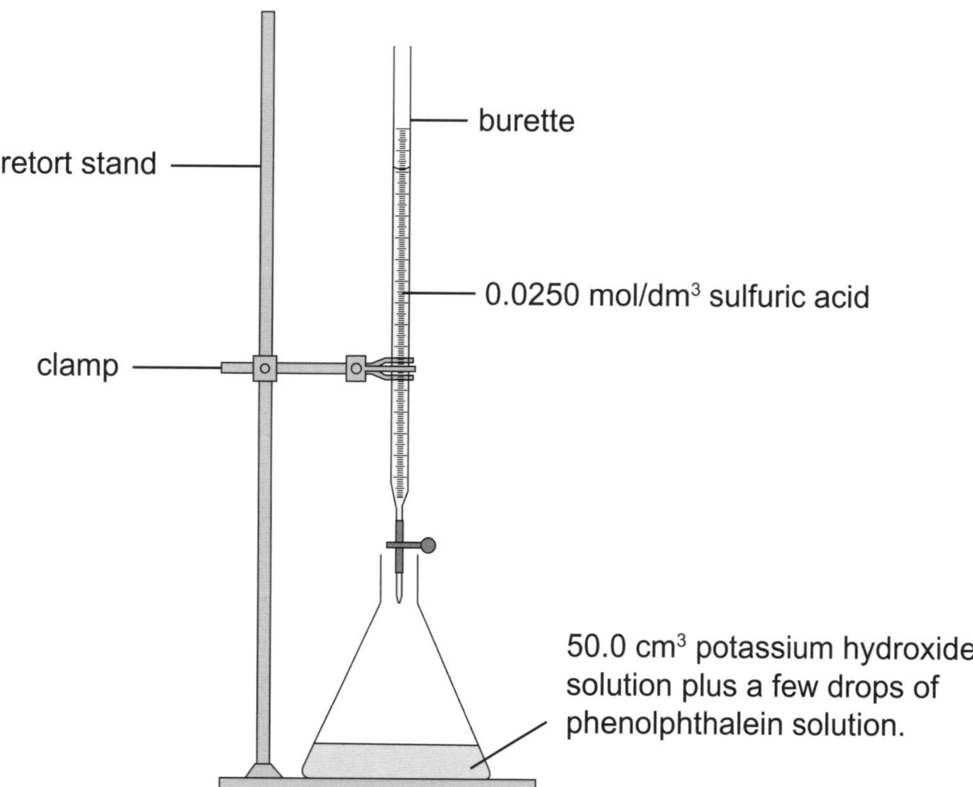

(a) Suggest **one** improvement to step 1.

....

....

(1)

(b) Suggest **one** safety precaution to take during step 4.

....

....

(1)

The student repeated the titration three times.

The results are shown in the table.

(c) Calculate the missing values.

Experiment	Initial reading / cm³	Final reading / cm³	Titre Volume / cm³
1	1.20	18.80	17.60
2	3.35	21.15	....
3	1.50	....	17.70

(2)

(d) Why did the student repeat the experiment?

....

....

(1)

The equation for the reaction between sulfuric acid and potassium hydroxide is shown below.

$$H_2SO_{4\,(aq)} + 2KOH_{(aq)} \rightarrow K_2SO_{4\,(aq)} + 2H_2O_{(l)}$$

(e) A second student performed the same titration.

Their average titre volume was 17.60 cm³.

How many moles of sulfuric acid reacted with the potassium hydroxide in this titration?

Number of moles = ...

(2)

12

(f)   How many moles of potassium hydroxide reacted with the acid in this titration?

Number of moles = ...
(1)

(g) Calculate the concentration of the potassium hydroxide solution in mol/dm^3.

Concentration = ... mol/dm^3
(2)

(Total for Question 6 = 10 marks)

7   Ethanol is a useful organic chemical.

(a) Ethanol contains a specific functional group.

(i)   Give the formula of the functional group in ethanol.

....
(1)

(ii) Ethanol is commonly produced by the fermentation of sugars at 30 °C.

Explain why increasing the temperature of the reaction would not increase
the rate of fermentation.

....

....

....
(2)

(iii) What type of reaction occurs when ethanol is burned in air?

....
(1)

(iv) Give the products of this reaction.

....
(1)

(b) Ethanol can be processed further to produce ethanoic acid.

(i)   Draw the displayed formula of ethanoic acid.

(1)

(ii) Give **one** way that ethanoic acid can be made from ethanol.

....

....
(1)

(iii) Calculate the percentage mass of oxygen in ethanoic acid.

% mass of oxygen = ... %

(2)

(Total for Question 7 = 9 marks)

8 Nitrogen can react with hydrogen to produce ammonia.

$$N_{2\,(g)} + 3H_{2\,(g)} \rightarrow 2NH_{3\,(g)}$$

(a) Bond energies for some of the bonds in the reaction as well as the overall enthalpy change for the reaction are shown below.

H-H = 436 kJ/mol   N-H = 391 kJ/mol   ΔH = –97 kJ/mol

(i) Use the data to calculate the bond energy for a N≡N bond.

(4)

(ii) Explain what would happen to the temperature of the surroundings during the reaction.

....

....

....

(2)

(b) In a second reaction, more energy is released by forming bonds than is used to break bonds.

Sketch an energy profile diagram for this reaction.
Label the products, reactants, activation energy and energy released.

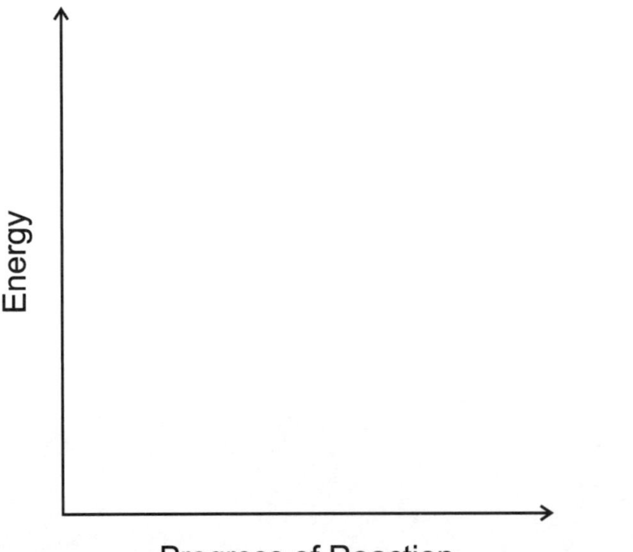

(2)

(Total for Question 8 = 8 marks)

9 Solid substances vary in whether they dissolve in certain liquids.

(a) The table below contains a list of different salts.
Complete the table to show whether each salt is soluble or insoluble in water.

Salts	**Soluble or insoluble in water?**
Calcium nitrate	
Lead(II) sulfate	
Sodium hydroxide	
Ammonium chloride	

(2)

(b) Sodium carbonate and copper carbonate are added to the same beaker of water.
A solution forms.

Identify the solute in the solution.

....

(1)

(c) What is meant by a saturated solution?

....

....

(1)

(d) A student runs an experiment to measure the solubility of a compound. Their data is shown in the table below.

Temperature / °C	Solubility / g per 100 g of solvent
0	5
20	12
40	23
60	40
80	69
100	130

(i) Plot a solubility curve of this data on the grid below.

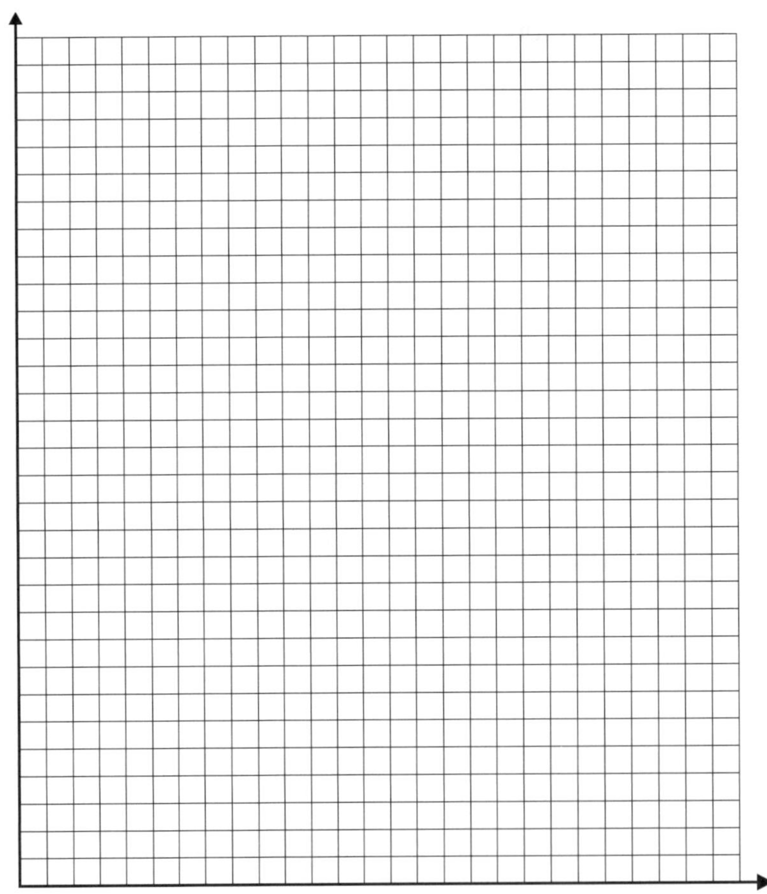

(3)

(ii) At what temperature would 80 g of the compound dissolve in 100 g of solvent?

....

(1)

(iii) The student then repeated the experiment using twice as much solvent. How would you expect the solubility to be affected? Explain your answer.

....

....

....

....

(2)

(Total for Question 9 = 10 marks)

TOTAL FOR PAPER = 70 MARKS

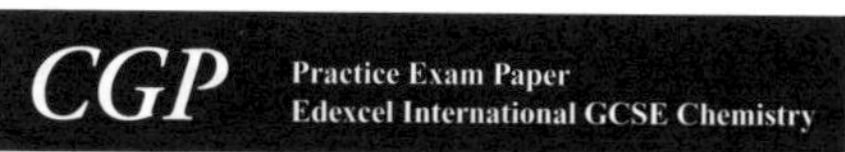

International GCSE Chemistry

Set B Paper 1

In addition to this paper you should have:
- A ruler.
- A calculator.

Centre name			
Centre number			
Candidate number			

Surname	
Other names	
Candidate signature	

Time allowed:
- 2 hours

Instructions to candidates
- Write your name and other details in the spaces provided above.
- Use a pen with black ink.
- You are allowed to use a calculator.
- Answer **all** questions in the spaces provided.
  There might be more space than you need.
- Answer multiple choice questions by putting a cross in the correct box.
  If you need to change your answer, draw a horizontal line through the box.
  Then mark your new answer as normal.

Information for candidates
- The marks available are given in brackets at the end of each question.
- There are 110 marks available for this paper.

Advice to candidates
- Try to answer all the questions.
- Carefully read each question before you try to answer it.
- If you have time at the end of the exam, check your answers.

For examiner's use

Q	Attempt Nº 1	Attempt Nº 2	Attempt Nº 3	Q	Attempt Nº 1	Attempt Nº 2	Attempt Nº 3
1				6			
2				7			
3				8			
4				9			
5				10			
Total							

The Periodic Table

	1 H Hydrogen 1																	Group 0

Relative atomic mass →
Atomic (proton) number →

Group 1	Group 2											Group 3	Group 4	Group 5	Group 6	Group 7	Group 0
																	4 He Helium 2
7 Li Lithium 3	9 Be Beryllium 4											11 B Boron 5	12 C Carbon 6	14 N Nitrogen 7	16 O Oxygen 8	19 F Fluorine 9	20 Ne Neon 10
23 Na Sodium 11	24 Mg Magnesium 12											27 Al Aluminium 13	28 Si Silicon 14	31 P Phosphorus 15	32 S Sulfur 16	35.5 Cl Chlorine 17	40 Ar Argon 18
39 K Potassium 19	40 Ca Calcium 20	45 Sc Scandium 21	48 Ti Titanium 22	51 V Vanadium 23	52 Cr Chromium 24	55 Mn Manganese 25	56 Fe Iron 26	59 Co Cobalt 27	59 Ni Nickel 28	63.5 Cu Copper 29	65 Zn Zinc 30	70 Ga Gallium 31	73 Ge Germanium 32	75 As Arsenic 33	79 Se Selenium 34	80 Br Bromine 35	84 Kr Krypton 36
85 Rb Rubidium 37	88 Sr Strontium 38	89 Y Yttrium 39	91 Zr Zirconium 40	93 Nb Niobium 41	96 Mo Molybdenum 42	98 Tc Technetium 43	101 Ru Ruthenium 44	103 Rh Rhodium 45	106 Pd Palladium 46	108 Ag Silver 47	112 Cd Cadmium 48	115 In Indium 49	119 Sn Tin 50	122 Sb Antimony 51	128 Te Tellurium 52	127 I Iodine 53	131 Xe Xenon 54
133 Cs Caesium 55	137 Ba Barium 56	139 La Lanthanum 57	178 Hf Hafnium 72	181 Ta Tantalum 73	184 W Tungsten 74	186 Re Rhenium 75	190 Os Osmium 76	192 Ir Iridium 77	195 Pt Platinum 78	197 Au Gold 79	201 Hg Mercury 80	204 Tl Thallium 81	207 Pb Lead 82	209 Bi Bismuth 83	209 Po Polonium 84	210 At Astatine 85	222 Rn Radon 86
223 Fr Francium 87	226 Ra Radium 88	227 Ac Actinium 89	261 Rf Rutherfordium 104	262 Db Dubnium 105	266 Sg Seaborgium 106	264 Bh Bohrium 107	277 Hs Hassium 108	268 Mt Meitnerium 109	271 Ds Darmstadtium 110	272 Rg Roentgenium 111							

The Lanthanides (atomic numbers 58-71) and the Actinides (atomic numbers 90-103) are not shown in this table.

1   This question is about hydrocarbons.

(a) Define the term **hydrocarbon**.

....

....

(2)

(b) Hydrocarbons are found in crude oil.

Name the process that is used to separate the hydrocarbons in crude oil.

....

(1)

(c) What property of hydrocarbons allows them to be separated?

☐ **A**   Electrical conductivity

☐ **B**   Boiling point

☐ **C**   Reactivity

☐ **D**   Viscosity

(1)

(d) Which of these crude oil fractions would be drained out furthest up
the fractionating column?

☐ **A**   Fuel oil

☐ **B**   Kerosene

☐ **C**   Refinery gases

☐ **D**   Gasoline (petrol)

(1)

(e) Which of these statements is **true**?

☐ **A**   Shorter hydrocarbons have a lower boiling point and are
more viscous.

☐ **B**   Shorter hydrocarbons have a higher boiling point and are less viscous.

☐ **C**   Shorter hydrocarbons have a lower boiling point and are less viscous.

☐ **D**   Shorter hydrocarbons have a higher boiling point and are
more viscous.

(1)

(f) Burning hydrocarbons releases gases into the atmosphere.

   (i) What are the products of incomplete combustion of hydrocarbons?

☐ **A** Carbon dioxide and carbon monoxide

☐ **B** Carbon dioxide, water and soot

☐ **C** Soot, oxygen and carbon monoxide

☐ **D** Carbon dioxide, carbon monoxide, water and soot

(1)

  (ii) Complete this table which shows gases and their approximate percentages in dry air.

Gas	Approximate percentage
Nitrogen	
	21%
	1%
Carbon dioxide	

(2)

(Total for Question 1 = 9 marks)

2 This question is about atoms.

(a) Atoms are made up of different subatomic particles.

(i) Describe the structure of an atom.

....

....

....
(2)

(ii) Complete the table with the masses and charges of the subatomic particles.

Particle	Relative mass	Relative charge
Proton		+1
Neutron		
Electron	0.0005	

(2)

(b) Complete the table with the number of protons, neutrons and electrons in an atom of the element with the symbol $^{7}_{3}Li$.

Particle	Number in atom
Proton	
Neutron	
Electron	

(1)

(c) Many elements can exist as various different isotopes.

(i) Explain which of the following are isotopes of each other.

$^{40}_{20}Ca$ $\quad$ $^{40}_{19}K$ $\quad$ $^{40}_{18}Ar$ $\quad$ $^{12}_{6}C$ $\quad$ $^{4}_{2}He$ $\quad$ $^{48}_{20}Ca$

....

....
(1)

(ii) The natural abundances of lithium-6 and lithium-7 are shown in the table.

	lithium-6	lithium-7
Percentage abundance	7.6	92.4

Use the information in the table to calculate the relative atomic mass of lithium, Li.

Give your answer to 2 significant figures.

Relative atomic mass = ...

(3)

(Total for Question 2 = 9 marks)

3 This question is about separating mixtures.

(a) Describe the difference between a mixture and a compound in terms of the bonding between components.

....

....

....

(2)

The table shows some information about four different substances.

Substance	Solubility in water	Melting point / °C	Boiling point / °C
Copper sulfate	Soluble	200	650
Sulfur	Not soluble	115	445
Water		0	100
Propanone	Soluble	−95	56

(b) Name a technique that could be used to separate the following mixtures.

Give a reason for your answer in each case.

Propanone and water: ....

....

....

Sulfur and water: ....

....

....

(4)

(c) Describe a method that could be used to separate a mixture of sulfur
and copper sulfate.

....

....

....

....

....

....

....

....

....

....

....

....

....

....

(6)

(Total for Question 3 = 12 marks)

4 Substances can be held together by ionic, covalent or metallic bonds.

The properties of a substance are determined by its bonding.

Some information about three different substances, **A**, **B** and **C**, is shown in the table.

Substance	Melting point / °C	Bonding	Electrical conductivity	
			As a solid	As a liquid
A	801	Ionic	Very low	High
B	−182	Covalent	Very low	Very low
C	1085	Metallic	High	High

(a) The three substances are copper, methane and sodium chloride.
Complete the table below by matching each substance to its identity.

Substance	Identity
A	
B	
C	

(2)

(b) Sodium chloride can be made from its elements, sodium and chlorine.

Describe, in terms of electrons, what happens to the sodium and chlorine atoms when they react to form sodium chloride.

....

....

....

(2)

(c) Explain why ionic substances have high melting points.

....

....

....

(2)

(d) Describe the difference in melting points between a substance with a giant covalent structure and a substance made of small molecules.

Explain your answer.

....

....

....

....

....

....

(3)

(e) A student reacts Substance **C** with oxygen gas to form an oxide as the only product.

(i) The mass of oxygen gas used in the reaction is measured to be 2.88 g.

Calculate the number of moles of oxygen gas used in the reaction.

(2)

(ii) Predict what type of reaction (if any) will occur when the oxide of Substance **C** is added to a solution of nitric acid.

Explain your answer.

....

....

....

(2)

(Total for Question 4 = 13 marks)

5 A student is investigating whether certain food colourings are pure chemicals or a mixture of different coloured chemicals.

He has a sample of blue food colouring and a sample of red food colouring.

The figure shows the apparatus the student used to analyse the blue food colouring.

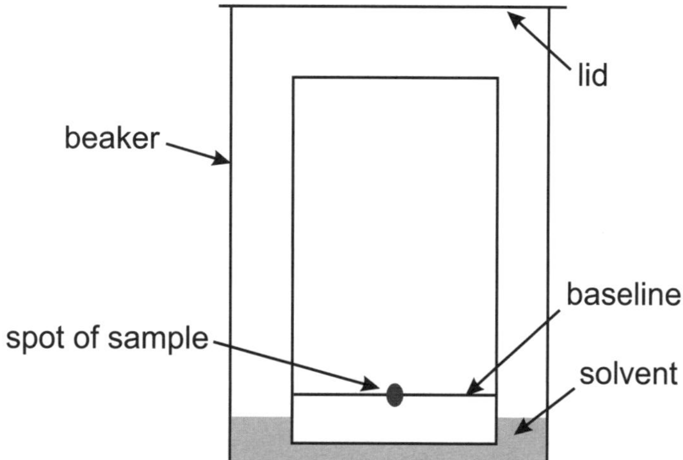

(a) The student drew the baseline in ink.

   (i) Why should the student have used a pencil?

....

(1)

The student starts the experiment again, but uses a piece of chromatography paper with a baseline drawn in pencil. One spot is present on the final chromatogram.

The student says that this means the sample is pure.

   (ii) Suggest why the student may not be correct.

....

....

(1)

   (iii) Explain why the student put a lid on the beaker.

....

....

(1)

The red food colouring is a mixture of four coloured chemicals. Each of the chemicals has a different distribution between the solvent and the filter paper.

(iv) Sketch a diagram to show what the chromatogram may look like at the end of the experiment.

(2)

(b) In another chromatography experiment, green, yellow, red and blue fabric dyes were analysed.
The figure shows the paper chromatogram that was produced.

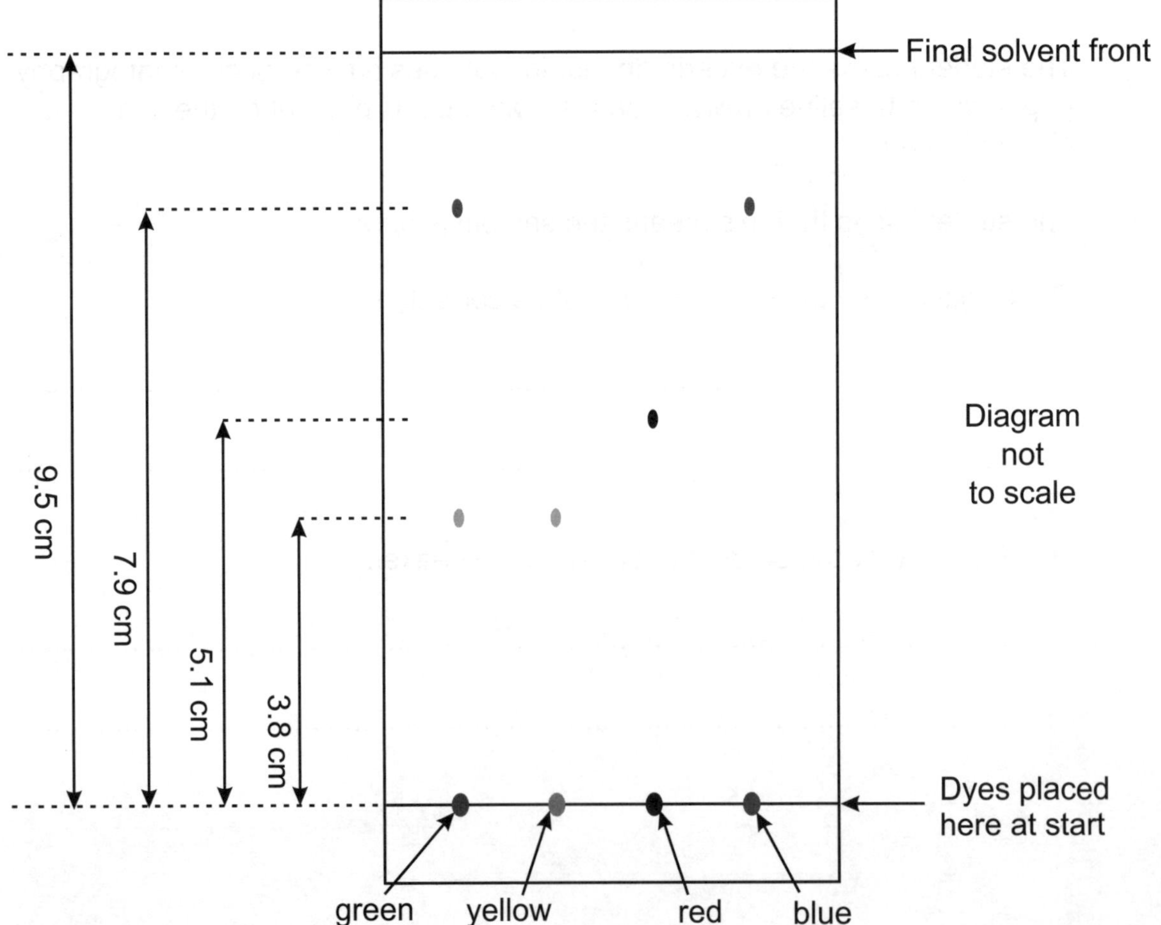

(i)   Calculate the R_f value for the blue dye.

Give your answer to 2 significant figures.

Blue R_f = ...

(3)

(ii) The green dye is made up of a mixture of dyes.

Which dyes does the chromatogram suggest are mixed to make the green dye?

Explain your reasoning.

....

....

....

(2)

(Total for Question 5 = 10 marks)

6 The graph shows the results of a study showing how the number of new vehicle registrations and the emissions of nitrogen oxides in **Country A** changed between 2010 and 2020.

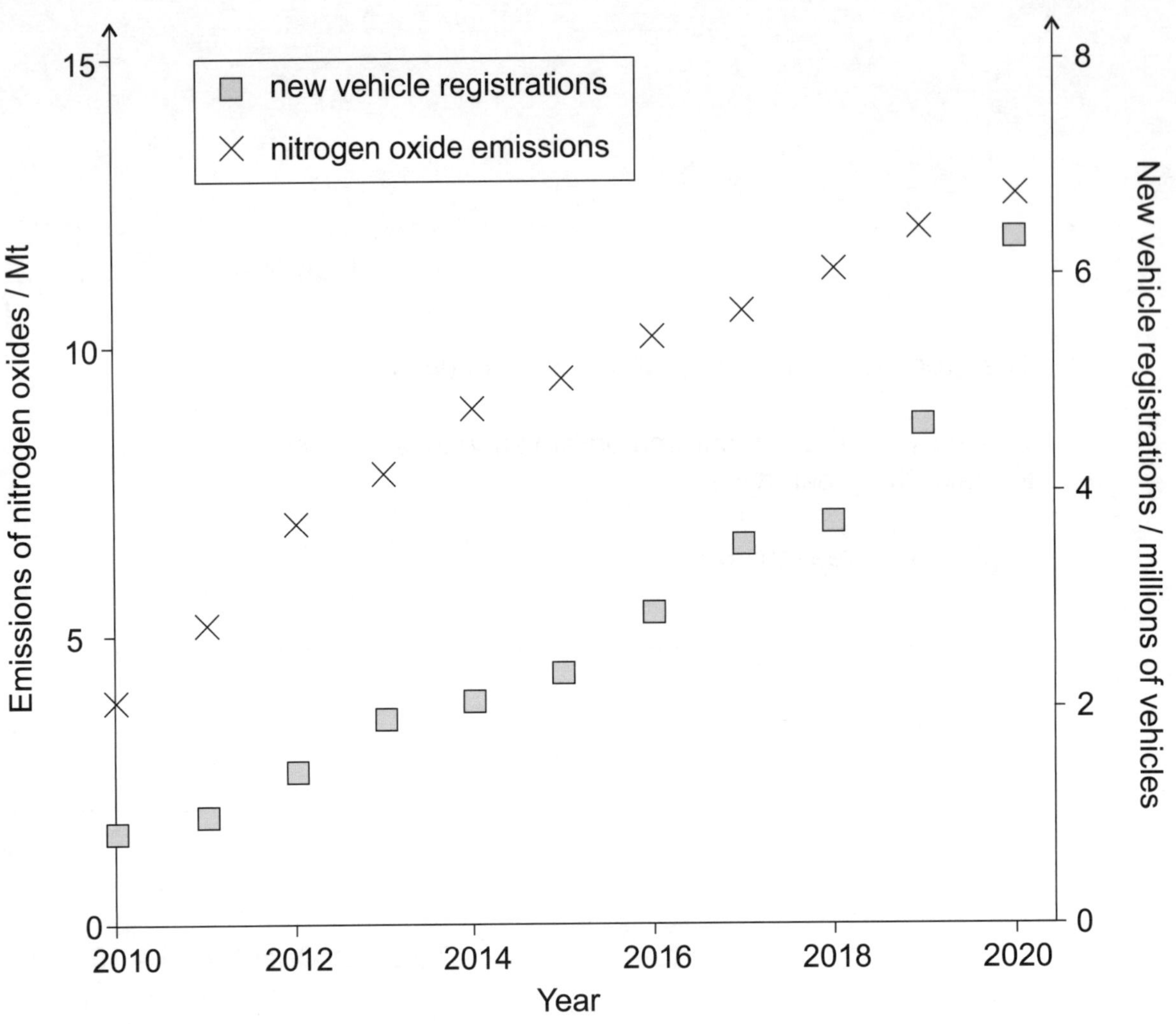

(a) Suggest an explanation for the trend in emissions of nitrogen oxides shown in the graph.

....

....

....

....

....

(3)

In 2018, 11.3 megatonnes (Mt) of nitrogen oxides were emitted.

A study predicts that emissions will be 14.0% higher in 2024 than they were in 2018.

(b) Calculate the mass of nitrogen oxides predicted to be emitted in 2024.

Give your answer to an appropriate number of significant figures.

Mass = ... Mt

(2)

(c) Using the data in the graph, suggest how the acidity of lakes in **Country A** might change between 2010 and 2024.

Explain your answer.

....

....

....

(2)

(d) Many fuels used in cars are low sulfur fuels.

Explain **one** benefit of reducing the amount of sulfur in fuels.

....

....

....

....

(3)

(Total for Question 6 = 10 marks)

7   Magnesium sulfate can be prepared and purified in the lab by following the method below.

1.   Slowly add magnesium powder to sulfuric acid, with stirring and gentle heating.

2.   Keep adding magnesium powder until there is excess solid left.

3.   Filter the mixture, boil the filtrate to reduce the volume by half and then leave to cool.

(a) It is important to follow the steps of this method carefully.

(i)   Explain why the mixture is heated.

....

....
(1)

(ii) Explain why magnesium powder is added in excess.

....

....
(1)

(iii) Explain what happens when the filtrate is left to cool.

....

....
(1)

(b) Suggest a test that could be used to check that all the sulfuric acid had reacted.

Describe the result that you would expect if all the acid had reacted.

Test: ....

Result: ....

....
(2)

(c) What ion is produced by all acids in aqueous solution?

....
(1)

(d) This is an example of a redox reaction.
     State which species is being oxidised in the reaction.

     Give a reason for your answer.

     Oxidised species: ....

     Reason: ....

     ....

(2)

(e) Sulfuric acid also reacts with zinc.

     A student adds an equal volume of dilute sulfuric acid to two boiling tubes.
     They add magnesium to one tube and an equal mass of zinc to the other.
     The pieces of each metal are of equal size and shape.

     (i)   How would the two reactions differ?
           Explain your answer.

           ....

           ....

           ....

           ....

(2)

     (ii) Another student reacts 1.25 g of zinc with an excess of sulfuric acid.
           The reaction produces 1.03 g of zinc sulfate.
           The equation for the reaction is:

$$Zn + H_2SO_4 \rightarrow ZnSO_4 + H_2$$

           Calculate the percentage yield of zinc sulfate.

           $M_r(ZnSO_4) = 161$

           Percentage yield = ... %

(4)

(Total for Question 7 = 14 marks)

8 Citric acid can react with sodium hydrogen carbonate.

The equation for the reaction is:

citric acid + sodium hydrogen carbonate $\rightarrow$ sodium citrate + water + carbon dioxide

A student carries out an experiment to observe the temperature change of the reaction.

The temperature of the solution was recorded at regular intervals and the results are shown in the table.

Time / s	Temperature of the solution / °C
0	26.0
5	24.5
10	22.5
15	22.5
20	19.5
25	18.0
30	16.5

(a) At which time interval was an anomalous result recorded?

.... seconds

(1)

(b) Describe the energy transfer that occurs during the reaction.

Explain how you know this.

....

....

....

....

(2)

18

(c) Describe a method that the student could use to investigate how the
    concentration of citric acid affects the temperature change in this reaction.

    Details about appropriate safety precautions are not required.

....

....

....

....

....

....

....

....

....

....

....

....

(6)

(Total for Question 8 = 9 marks)

9 A student is carrying out tests to identify solutions of four different compounds.

(a) She carries out a flame test on Solution 1. She sees a lilac flame.

Identify the metal ion that is present in Solution 1.

....

(1)

(b) Solution 2 contains Fe^{2+} ions and Solution 3 contains Fe^{3+} ions.

Describe a test that the student could carry out to distinguish between these solutions. State the expected result for each solution.

....

....

....

....

....

....

(3)

(c) The student carries out three different tests on Solution 4.

Her results are shown in the table.

Test	Result
Flame test	Yellow flame
Add a few drops of hydrochloric acid followed by barium chloride solution.	Solution remains clear
Add a few drops of dilute nitric acid followed by silver nitrate solution.	Cream precipitate

(i) Why was hydrochloric acid added to Solution 4 before testing with barium chloride?

....

....

(1)

(ii) Suggest the identity of the compound in Solution 4.

Explain your answer.

....

....

....

....

....

....

....

(4)

(Total for Question 9 = 9 marks)

10 Fluorine, chlorine and bromine are elements known as halogens.
Halogens exist as covalent molecules with the general formula X_2.

(a) Predict what would you would observe if chlorine gas was bubbled through a
solution of potassium bromide (KBr).

Give a reason for your answer.

....

....

....

(2)

(b) 3.40 g of rubidium reacts with an excess of an unknown halogen to form
a single, ionic compound as the product.

The mass of the product is 4.82 g.

Which halogen was used in this reaction?

unknown halogen = ...

(4)

(c) Bromine reacts with copper to form copper(II) bromide.

(i) What is the formula for copper(II) bromide?

....

(1)

Bromine can also react with butene.

$$Br_2 + C_4H_8 \rightarrow C_4H_8Br_2$$

(ii) 3.36 g of bromine is reacted with 3.36 g of butene.

Which of the reactants is in excess?
Show your working.

reactant in excess = ...

(3)

(d) Chlorine can react with hydrocarbons, such as methane.

(i) Give the condition required for chlorine to react with methane.

....

....

(1)

(ii) Reacting chlorine and methane produces chloromethane.

Draw a dot and cross diagram for chloromethane.

(2)

Chlorine can also undergo reactions with propane.

(iii) Name two products with the molecular formula C_3H_7Cl that could be produced by propane's reaction with chlorine.

....

....

(2)

(Total for Question 10 = 15 marks)

TOTAL FOR PAPER = 110 MARKS

International GCSE Chemistry

Set B Paper 2

In addition to this paper you should have:
- A ruler.
- A calculator.

Centre name				
Centre number				
Candidate number				

Surname	
Other names	
Candidate signature	

Time allowed:
- 1 hour 15 minutes

Instructions to candidates
- Write your name and other details in the spaces provided above.
- Use a pen with black ink.
- You are allowed to use a calculator.
- Answer **all** questions in the spaces provided.
  There might be more space than you need.
- Answer multiple choice questions by putting a cross in the correct box.
  If you need to change your answer, draw a horizontal line through the box.
  Then mark your new answer as normal.

Information for candidates
- The marks available are given in brackets at the end
  of each question.
- There are 70 marks available for this paper.

Advice to candidates
- Try to answer all the questions.
- Carefully read each question before you try to answer it.
- If you have time at the end of the exam, check your answers.

	For examiner's use						
Q	Attempt Nº			Q	Attempt Nº		
	1	2	3		1	2	3
1				5			
2				6			
3				7			
4							
				Total			

The Periodic Table

Key:
- Relative atomic mass (top number)
- Atomic (proton) number (bottom number)

1 H Hydrogen 1

Group 1	Group 2											Group 3	Group 4	Group 5	Group 6	Group 7	Group 0
																	4 He Helium 2
7 Li Lithium 3	9 Be Beryllium 4											11 B Boron 5	12 C Carbon 6	14 N Nitrogen 7	16 O Oxygen 8	19 F Fluorine 9	20 Ne Neon 10
23 Na Sodium 11	24 Mg Magnesium 12											27 Al Aluminium 13	28 Si Silicon 14	31 P Phosphorus 15	32 S Sulfur 16	35.5 Cl Chlorine 17	40 Ar Argon 18
39 K Potassium 19	40 Ca Calcium 20	45 Sc Scandium 21	48 Ti Titanium 22	51 V Vanadium 23	52 Cr Chromium 24	55 Mn Manganese 25	56 Fe Iron 26	59 Co Cobalt 27	59 Ni Nickel 28	63.5 Cu Copper 29	65 Zn Zinc 30	70 Ga Gallium 31	73 Ge Germanium 32	75 As Arsenic 33	79 Se Selenium 34	80 Br Bromine 35	84 Kr Krypton 36
85 Rb Rubidium 37	88 Sr Strontium 38	89 Y Yttrium 39	91 Zr Zirconium 40	93 Nb Niobium 41	96 Mo Molybdenum 42	98 Tc Technetium 43	101 Ru Ruthenium 44	103 Rh Rhodium 45	106 Pd Palladium 46	108 Ag Silver 47	112 Cd Cadmium 48	115 In Indium 49	119 Sn Tin 50	122 Sb Antimony 51	128 Te Tellurium 52	127 I Iodine 53	131 Xe Xenon 54
133 Cs Caesium 55	137 Ba Barium 56	139 La Lanthanum 57	178 Hf Hafnium 72	181 Ta Tantalum 73	184 W Tungsten 74	186 Re Rhenium 75	190 Os Osmium 76	192 Ir Iridium 77	195 Pt Platinum 78	197 Au Gold 79	201 Hg Mercury 80	204 Tl Thallium 81	207 Pb Lead 82	209 Bi Bismuth 83	209 Po Polonium 84	210 At Astatine 85	222 Rn Radon 86
223 Fr Francium 87	226 Ra Radium 88	227 Ac Actinium 89	261 Rf Rutherfordium 104	262 Db Dubnium 105	266 Sg Seaborgium 106	264 Bh Bohrium 107	277 Hs Hassium 108	268 Mt Meitnerium 109	271 Ds Darmstadtium 110	272 Rg Roentgenium 111							

The Lanthanides (atomic numbers 58-71) and the Actinides (atomic numbers 90-103) are not shown in this table.

Answer all questions in the spaces provided

1 This question is about reversible reactions.

Sulfur trioxide is produced in a reversible reaction.

The equation for the reaction is:

$$\text{sulfur dioxide} + \text{oxygen} \rightleftharpoons \text{sulfur trioxide} \qquad \Delta H = -196 \text{ kJ/mol}$$

(a) What is the correct balanced equation for the reaction?

☐ **A** $\quad 2SO_2 + O_2 \rightleftharpoons SO_3$

☐ **B** $\quad SO_2 + 2O \rightleftharpoons SO_3$

☐ **C** $\quad 2SO_2 + O_2 \rightleftharpoons 2SO_3$

☐ **D** $\quad 2SO_2 + 2O_2 \rightleftharpoons 2SO_3$

(1)

(b) The reaction of sulfur dioxide and oxygen to form sulfur trioxide is exothermic.

Which of the following statements is true?

☐ **A** The activation energy for the reaction of sulfur dioxide with oxygen is negative.

☐ **B** Sulfur trioxide is colder than sulfur dioxide and oxygen.

☐ **C** Heating the reaction will cause more sulfur trioxide to form.

☐ **D** The reaction of sulfur trioxide to form sulfur dioxide and oxygen is endothermic.

(1)

(c) Which of these conditions is essential for a reversible reaction to reach equilibrium?

☐ **A** The reactants and products must not be able to escape from the reaction vessel.

☐ **B** The temperature and pressure of the reaction must not change.

☐ **C** The formation of products must be exothermic and the formation of reactants must be endothermic.

☐ **D** A catalyst must not be present in the reaction mixture.

(1)

(d) When a reversible reaction has reached equilibrium, which of the following is true?

    ☐ **A**   The forward and reverse reactions have both stopped happening.

    ☐ **B**   The rates of the forward and reverse reactions are equal.

    ☐ **C**   The amount of product will be the same as the amount of reactant.

    ☐ **D**   The temperature and pressure can no longer be changed.

(1)

(e) Sulfur dioxide, oxygen and sulfur trioxide are all gases.

During the reaction, the equilibrium shifts to increase the yield of sulfur trioxide.

Suggest what change in conditions may have led to this.
Justify your answer.

....

....

....

(2)

(Total for Question 1 = 6 marks)

2 The diagram shows how a student extracted copper from a copper ore by heating the ore with carbon in a crucible.

The copper ore is mainly copper carbonate.

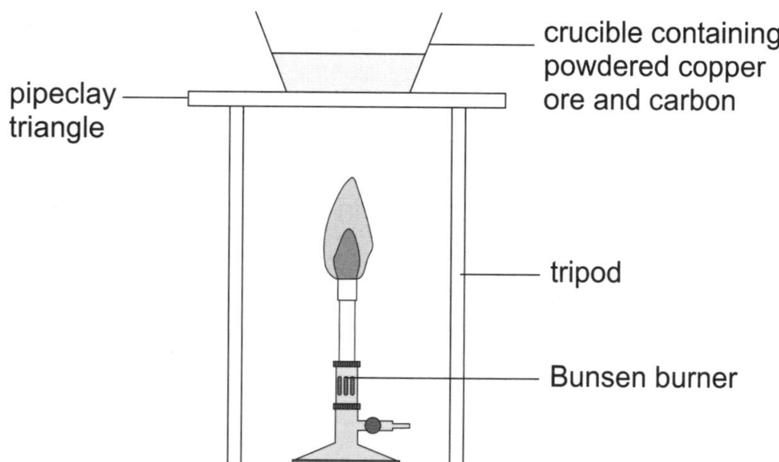

(a) Suggest **two** safety precautions the student should take when doing this experiment.

1. ....

2. ....

(2)

(b) Two reactions happen during the experiment.
In the first reaction, copper carbonate thermally decomposes to form copper oxide and carbon dioxide.

$$CuCO_3 \rightarrow CuO + CO_2$$

Calculate the mass of copper carbonate that must decompose to produce 0.050 moles of copper oxide.

Mass of copper carbonate = ... g

(2)

(c) In the second reaction, the copper oxide reacts with carbon to form copper and carbon dioxide.

$$2CuO + C \rightarrow 2Cu + CO_2$$

(i) State the type of reaction that the copper oxide undergoes in this reaction.

....
(1)

(ii) Explain why carbon is suitable for use in this reaction.

Give your answer in terms of reactivity.

....
(1)

(Total for Question 2 = 6 marks)

3   A student decided to investigate the conditions needed for iron to rust.

She stored some iron nails in different conditions for one week.

(a) The results of the experiment are shown in the table.

	Nail 1	**Nail 2**	**Nail 3**
Storage conditions	Stored in water open to the air.	Stored in dry air.	Stored in pure water with no air present.
Appearance after one week	Very rusty	No rust	No rust

(i)   What do the results show about the substances needed for rusting to occur?

....

....

(1)

The student then set up the experiment shown in the diagram.

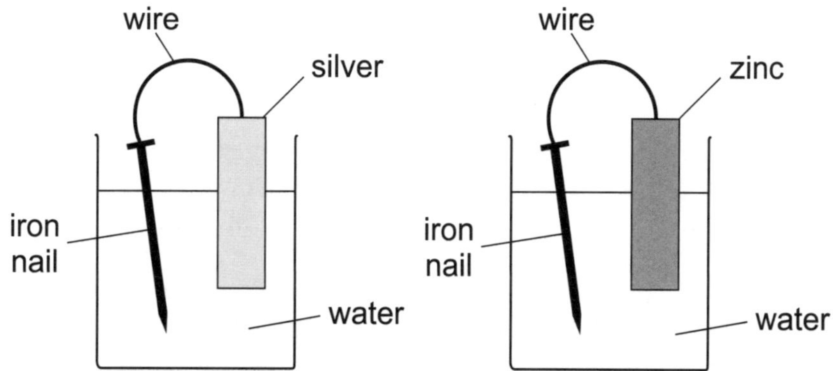

(ii) Explain which nail is likely to have no rust after one week.

....

....

....

(1)

(b) Steels are examples of alloys.

(i)   Why can steel be described as an iron alloy?

....

....

(1)

(ii) Why are pure metals often unsuitable for everyday use?

....

(1)

Different steels have different properties.

(iii) Give **one** use for stainless steel.

Explain why it is suitable for this use.

Use: ....

Explanation: ....

....

....

(2)

(Total for Question 3 = 6 marks)

8

4 It is possible to produce both insoluble and soluble salts in the laboratory.

(a) A student dissolves lead nitrate and magnesium sulfate in separate test tubes using deionised water. The student then tips both solutions into a beaker and stirs them together.

The student notices that an insoluble salt has precipitated out of the solution.

(i) Name this precipitate.

....

(1)

(ii) Why should deionised water be used to dissolve the reactants?

....

....

(1)

(iii) When conducting this experiment, the insoluble salt should be dried.

State **one** piece of equipment the student could use to dry the salt.

....

(1)

(b) Nitric acid can be reacted with calcium hydroxide to produce a soluble salt.

(i) Give the molecular formula of the salt produced by this reaction.

....

(1)

(ii) A student conducts this experiment using 450 cm³ of 1.5 mol/dm³ nitric acid.

How many moles of nitric acid are there in this solution?

Number of moles = ... mol

(2)

(c) When water is added to zinc nitrate, it forms a hydrated salt, $Zn(NO_3)_2.xH_2O$.

A student measured the mass of an empty crucible then added a sample of hydrated zinc nitrate and measured the mass again. The student heated the crucible until the mass remained constant, then measured the final mass of the crucible and sample.

The student's results are shown in the table.

mass of empty crucible	61.500 g
starting mass of crucible + sample	65.233 g
final mass of crucible + sample	63.876 g

Calculate the value of x in $Zn(NO_3)_2.xH_2O$.

$M_r(Zn(NO_3)_2) = 189$

$M_r(H_2O) = 18$

Value of x = ...

(4)

(Total for Question 4 = 10 marks)

5   A student has a sample of hydrochloric acid with an unknown concentration.

He plans to determine the concentration of the hydrochloric acid by carrying out a titration with potassium hydroxide, KOH.

Potassium hydroxide is an alkali.

(a) Define the term 'alkali'.

....

....

(1)

(b) The student has 10 cm^3 of a potassium hydroxide solution with a concentration of 3 mol/dm^3.

Calculate the volume of water that he needs to add to dilute the solution to a concentration of 1.2 mol/dm^3.

Volume of water = ... cm^3

(3)

(c) Describe a method that the student could use to carry out the titration.

....

....

....

....

....

....

....

....

....

....

....

....

....

....

(6)

0.025 dm³ of 1.2 mol/dm³ potassium hydroxide solution was titrated against the hydrochloric acid.

The experiment was repeated three times.

The results are shown in the table.

Repeat	Volume of hydrochloric acid / cm³
1	12.80
2	12.75
3	12.85

The equation for the reaction is:

$$HCl + KOH \rightarrow KCl + H_2O$$

(d) Calculate the concentration of the hydrochloric acid used in the experiment.

Give your answer to three significant figures.

Concentration of acid = ... mol/dm³

(4)

(Total for Question 5 = 14 marks)

6 This question is about carboxylic acids.

(a) CH_3COOH is the formula of a carboxylic acid.

(i) What is the name of the carboxylic acid with formula CH_3COOH?

....

(1)

(ii) CH_3COOH can be reacted with butanol to form an ester.

Draw the displayed formula of the ester that is formed.

(1)

(b) A student is investigating the reactions of carboxylic acids with sodium carbonate.

(i) Carbon dioxide gas is produced when methanoic acid reacts with sodium carbonate.

Describe a method that could be used to confirm that the gas produced is carbon dioxide.
Give the expected result if carbon dioxide is present.

Method: ....

....

Expected result if carbon dioxide is present: ....

....

(2)

The student wants to measure the rate of reaction between solutions of methanoic acid and sodium carbonate.

She bubbles the gas produced through a large beaker of water and counts how many bubbles are produced in one minute. The bubbles are produced very quickly.

(ii) Explain why the bubbles being produced quickly could be a problem.

....

(1)

(iii) Suggest **one** way in which the rate of bubbling could be made slower.

Explain, using collision theory, why this change would affect the rate of the reaction.

....

....

....

....

(3)

(iv) The student carries out her experiment several more times to check whether her results are repeatable.

How could she test the reproducibility of her data?

....

....

(1)

The table shows the data collected in one of the experiments.

Time / s	0	15	30	45	60	75	90
Volume of gas / cm³	0	10	20	29	35	38	38

(v) On the graph paper:

- plot this data on the graph,
- draw a line of best fit.

(4)

(vi) Use the graph from part (v) to calculate the rate of the reaction at 45
seconds.

Rate of reaction at 45 s = ... cm^3/s

(3)

(Total for Question 6 = 16 marks)

7 Some information about three different ionic compounds is shown in the table below.

Compound	Ions	Melting point / °C	Solubility in water
Lead bromide	Pb^{2+} and Br^-	373	Low
Potassium carbonate	K^+ and CO_3^{2-}	891	High
Potassium hydroxide	K^+ and OH^-	360	High

This table shows the order of reactivity for several elements:

Element	Order of reactivity
Potassium	Most reactive
Sodium	
Aluminium	
Hydrogen	
Silver	
Gold	Least reactive

(a) Before electrolysis is carried out, solid lead bromide is melted and solid potassium carbonate is dissolved in water.

State the purpose of melting and dissolving these compounds before electrolysis.

....

....

(1)

(b) Electrolysis is a redox process.

At which electrode does oxidation occur during electrolysis?

Explain your answer.

Electrode: ....

Explanation: ....

....

....

(3)

Potassium metal can be obtained by electrolysing molten potassium carbonate or molten potassium hydroxide.

(c) Explain why potassium hydroxide is the preferred source of potassium by electrolysis.

....

....

....

(1)

(d) A student carries out electrolysis on an aqueous solution of one of the compounds in the table.

(i)   Explain why platinum would be a good choice of material for the electrodes.

....

(1)

The half equation for the reaction that occurs at one of the electrodes is:

$$4OH^- \rightarrow O_2 + 2H_2O + 4e^-$$

(ii) Write the half equation that occurs at the other electrode.

Include state symbols in your answer.

....

(2)

(iii) The student measures that 4.8 g of oxygen gas was given off during the reaction.

Calculate the volume of oxygen gas given off in cm^3.
The molar volume of oxygen gas at RTP is 24 dm^3.

Volume of oxygen = ... cm^3

(2)

(iv) How could the student confirm that the gas given off is oxygen?

....

....

....

(2)

(Total for Question 7 = 12 marks)

TOTAL FOR PAPER = 70 MARKS

Edexcel International GCSE

Chemistry

For the Grade 9-1 Course

Practice Exam Papers
Instructions & Answer Book

Perfect exam practice from CGP!

You can't bluff your way through Edexcel's International GCSE Chemistry exams. No chance. What you need is a way of making sure you're 100% prepared.

That's where this brilliant pack from CGP comes in. It contains two full sets of realistic mock exams, so you get used to tackling the types of questions examiners love to ask — all in the comfort of your own home / classroom / private jet.

We've also included full answers and mark schemes for each paper, so it's easy to check how you're getting on. You'll be ready for anything when the real exams roll around.

What to Expect in The Exams

(1) There will be Two Papers

For Edexcel International GCSE Chemistry, you'll sit <u>two exam papers</u> at the <u>end</u> of your course.

Paper	Time	No. of marks
1	2 hours	110
2	1 hr 15 mins	70

Some of the course content will only be assessed in Paper 2.

(2) You'll be Tested on your Maths...

Some of the marks for Edexcel International GCSE Chemistry come from questions on the <u>maths skills</u> you've used in the course. You'll be expected to remember the formulas for questions like mole calculations, so make sure you know them for your exam.

(3) ...and on your Practical Skills

- Edexcel International GCSE Chemistry contains <u>twelve practical investigations</u> that you'll do during the course — but you can also be asked about them in the exams.
- For example, you might be asked to comment on the <u>design</u> of an experiment (the <u>apparatus</u> and <u>methods</u>), make <u>predictions</u>, <u>analyse</u> or <u>interpret results</u>... Pretty much anything to do with planning and carrying out the investigations.

You could be asked about practical investigations you're unfamiliar with too. So you'll need to be able to apply the skills you've learnt to other experiments.

Marking Your Papers

- Do a complete exam set (Paper 1 and Paper 2).

- Use the answers and mark scheme in this booklet to mark each exam paper.

- Write down your mark for each paper in the table below —
  Paper 1 is worth 110 marks and Paper 2 is worth 70 marks.

- Find your total for the whole exam (out of a maximum of 180 marks)
  by adding up your marks from both papers.

- Follow the instructions below to estimate your grade.

	Paper 1	Paper 2	Total mark	Grade
SET A				
SET B				

Estimating Your Grade

- If you want to get a **rough idea** of the grade you're working at, we suggest you compare the **total mark** you got in **each set** to the latest set of grade boundaries.

- Grade boundaries are set for each individual exam, so they're likely to **change** from year to year. You can find the latest set of grade boundaries by going to **www.cgpbooks.co.uk/gcsegradeboundaries**

- Jot down the marks required for each grade in the table below so you don't have to refer back to the website. Use these marks to **estimate your grade**. If you're borderline, don't push yourself up a grade — the real examiners won't.

Total mark required for each grade									
Grade	9	8	7	6	5	4	3	2	1
Total mark out of 180									

- Remember, this will only be a **rough guide**, and grade boundaries will be different for different exams, but it should help you to see how you're getting on.

Published by CGP

Editors: Mary Falkner, Rob Hayman and Sharon Keeley-Holden.
Contributor: Chris Workman
Proofreaders: Emma Clayton, Katie Fernandez and Jamie Sinclair.

Many thanks to Jan Greenway for the copyright research.

Clipart from Corel®
Illustrations by: Sandy Gardner Artist, email sandy@sandygardner.co.uk
Printed by Elanders Ltd, Newcastle upon Tyne.
Text, design, layout and original illustrations
© Coordination Group Publications Ltd. (CGP) 2020
All rights reserved.

Answers

Set A — Paper 1

1 a)   **B** *[1 mark]*
Elements in the same group have the same number of electrons in their outer shell.

   b)   **C** *[1 mark]*
The Group 0 elements, or the noble gases, all have eight electrons in their outer shell (except helium, which has two).

   c)   **A** *[1 mark]*
The element labelled 3 in the table is in Group 7. Group 7 elements gain one electron to form −1 ions.

   d)   **C** *[1 mark]*
It takes a lot of energy to add or remove electrons from Group 0 atoms, so they're very unreactive.

   e)   2, 8, 8, 1 *[1 mark]*

   f)   Any three from: 2, 4, 5, 7, 8 *[1 mark]*

2 a) i) $ZnO + \mathbf{2HNO_3} \rightarrow Zn(NO_3)_2 + \mathbf{H_2O}$
    [1 mark for the reactant and product correct, 1 mark for correct balancing]

   ii) neutralisation *[1 mark]*

   iii) The student will know the reaction is complete when no more zinc oxide can be dissolved in the acid / when the excess zinc oxide sinks to the bottom of the flask and stays there *[1 mark]*.

   iv) Gently heat the zinc nitrate solution using a Bunsen burner/water bath/electric heater *[1 mark]* to evaporate some of the water *[1 mark]*. Then stop heating it and leave the solution to cool *[1 mark]*. Filter out the crystals that form and leave them to dry *[1 mark]*.

  b) i) Relative formula mass of $CuO = 63.5 + 16 = 79.5$
    Relative formula mass of $CuCl_2 = 63.5 + (35.5 \times 2) = 134.5$
    Moles of $CuO = $ mass $\div M_r = 10.1 \div 79.5 = 0.1270...$
    1 mole of CuO reacts to produce 1 mole of $CuCl_2$, so
    mass $= $ moles $\times M_r = 0.1270... \times 134.5 = \mathbf{17.1\ g\ of\ CuCl_2}$
    [3 marks for correct answer, otherwise 1 mark for correctly calculating both formula masses and 1 mark for calculating the number of moles of $CuCl_2$]

   ii) Percentage yield = (mass of product actually made ÷ maximum theoretical mass of product) × 100
    $(8.4 \div 12) \times 100 = \mathbf{70\%}$ *[1 mark]*

3 a) i) C_{60} fullerenes are hollow spheres made up of 60 carbon atoms *[1 mark]*. These carbon atoms are joined together by covalent bonds *[1 mark]*.

   ii) C_{60} molecules are large *[1 mark]*. This means the intermolecular forces between the molecules are relatively strong *[1 mark]*. This means a lot of energy would be required to break these forces and melt C_{60} *[1 mark]*.

  b) Graphite consists of layers of covalently bonded carbon atoms *[1 mark]*. These layers are only held together by weak intermolecular forces so can slide over each other *[1 mark]*.

  c) i) isotopes *[1 mark]*

   ii) C is 1% ^{13}C and 99% ^{12}C *[1 mark]*.
    $A_r = ((13 \times 1) + (12 \times 99)) \div 100$
    $= (13 + 1188) \div 100 = 12.01 = \mathbf{12}$ *[1 mark]*

4 a) alkanes *[1 mark]*

  b) $C_{10}H_{22} \rightarrow C_7H_{16} + \mathbf{C_3H_6}$ *[1 mark]*

  c) The decane is heated until it vaporises / turns into a gas *[1 mark]*. The vapour is passed over a silica or alumina catalyst at 600-700 °C *[1 mark]*. This cracks the decane into small alkanes (such as propane) and alkenes *[1 mark]*.

  d) Bromine water can be shaken with samples of the molecules *[1 mark]*. When bromine water is shaken with an alkane, it will stay orange *[1 mark]*. When bromine water is shaken with an alkene, it will decolourise/turn colourless *[1 mark]*.

  e) The supply of kerosene is higher than the demand *[1 mark]*. Cracking kerosene produces shorter-chain hydrocarbons, such as those found in the gasoline and refinery gases fractions *[1 mark]*. The demand for these fractions exceeds the supply *[1 mark]*.

5 a) i)

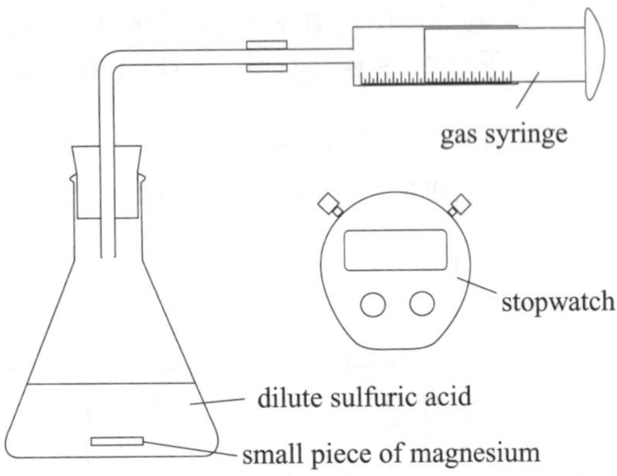

    [3 marks — 1 mark for dilute sulfuric acid and small piece of magnesium in a conical flask, 1 mark showing a suitable way of collecting all of the gas produced, and 1 mark for a stopwatch]
You'd also get the marks if you drew a set-up which involved collecting the gas over water — for example, using an upturned measuring cylinder in a beaker of water.

   ii) E.g. Hydrogen has a low molecular mass *[1 mark]*, so the change in mass as hydrogen leaves the reaction vessel would be very small and therefore difficult to measure *[1 mark]*.

  b) i) Volume of gas produced at 20 seconds = 100 cm³
    Volume of gas produced at 70 seconds = 145 cm³
    145 cm³ − 100 cm³ = **45 cm³** *[1 mark]*

   ii)

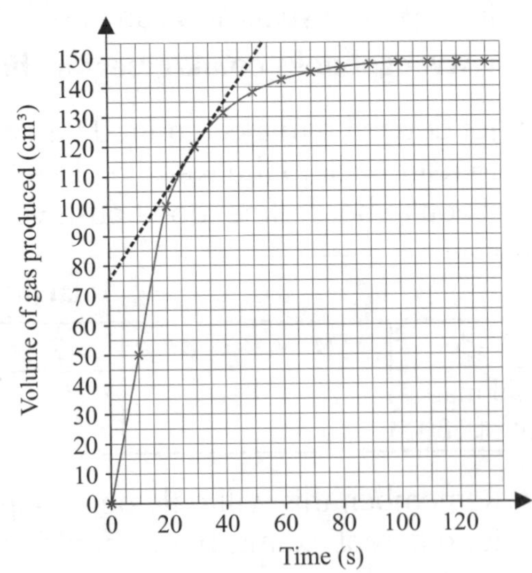

    E.g. Change in volume = 150 − 90 = 60 cm³
    Change in time = 50 − 10 = 40 s
    Rate = 70 ÷ 50 = **1.5 cm³/s**
    [Answers between 1.2 and 1.8 cm³/s are allowed. 3 marks for the correct answer, otherwise 1 mark for drawing a suitable tangent and 1 mark for calculating the change in volume and time]
Don't worry if your tangent isn't in exactly the same place or if you have got slightly different numbers. As long as your tangent was sensible, your final answer should be within the range of acceptable answers.

   iii) The rate of reaction would decrease *[1 mark]*. Lowering the concentration of the acid reduces the number of particles of acid present in the reaction *[1 mark]*, so collisions would be less frequent *[1 mark]*.

iv) E.g.

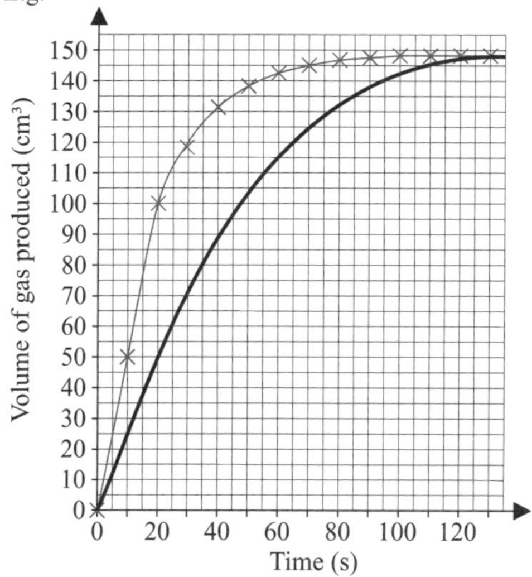

*[1 mark for a less-steep curve, 1 mark for finishing
at the same height but taking longer to reach it]*

6 a) i) A blue precipitate forms *[1 mark]*.
   ii) $CuCl_2 + 2NaOH \rightarrow Cu(OH)_2 + 2NaCl$ *[1 mark for correct
      formulas, 1 mark for correctly balanced equation]*
   b) You can test for water using white anhydrous
      copper(II) sulfate powder *[1 mark]*. To test the solution,
      add a couple of drops to the powder *[1 mark]*. If water
      is present, the white powder will turn blue *[1 mark]*.
   c) i) The student hasn't cleaned the platinum loop *[1 mark]*.
      This means that the flame colour could be affected by
      impurities *[1 mark]*.
      ii) lithium *[1 mark]* sulfate *[1 mark]*
7 a) E.g. the student could use a different solvent /
      The student could use a longer piece of chromatography
      paper and run the experiment for longer *[1 mark]*.
*The R_f values of the compounds will be different in different solvents,
so they may separate better if the solvent is changed. Running the
experiment for longer will give the chemicals more time to separate.*
   b) Distance travelled by solvent = 8.0 cm
      R_f = distance travelled by substance ÷
      distance travelled by solvent
      Distance travelled by substance =
      R_f × distance travelled by solvent = 0.60 × 8.0 = 4.8 cm
      chlorophyll a = spot C
      *[3 marks for correct answer, otherwise 1 mark for
      correct measurement of solvent line, 1 mark for using the
      correct equation]*
*You could also answer this question by working out the R_f values of each
spot on the chromatogram.*
   c) E.g. the student could measure the melting point of the
      crystals *[1 mark]*, and see whether it matches the reference
      melting point for chlorophyll a shown in data books /
      whether it has a single sharp melting point, as opposed to a
      range *[1 mark]*.
      Or
      The student could run a number of chromatography
      experiments on the crystals in different solvents *[1 mark]*.
      If the experiments always show one spot, then the crystals
      are pure *[1 mark]*.
   d) Distance travelled by spot A in experiment 1 = 1.3 cm
      R_f of spot A = distance travelled by substance ÷ distance
      travelled by solvent = 1.3 ÷ 8.0 = 0.1625
      Distance travelled by spot A in experiment 2 =
      R_f × distance travelled by solvent
      = 0.1625 × 12 = 1.95 cm = **2.0 cm**
      *[3 marks for correct answer, otherwise 1 mark for correct
      measurement of spot A on diagram, 1 mark for correctly
      calculating R_f of spot A]*

8 a) i) addition reaction *[1 mark]*
*There are other ways to classify reactions, so any other correct
classification would get you the mark.*
      ii) Propene is unsaturated/contains a C=C double bond
         [1 mark].
      iii)
      [structural formula of propane]
      [1 mark]
      iv) (3 × 12) + (6 × 1) + (2 × 80) = **202** *[1 mark]*
      v)
      [structural formula showing repeating unit with CH₃]
      [1 mark for correct formula, 1 mark for correct bonds]
      vi) E.g. they can release toxic gases *[1 mark]*.
   b) i) pent-2-ene *[1 mark]*
      ii) $2C_5H_{10} + 15O_2 \rightarrow 10CO_2 + 10H_2O$
         *[2 marks — 1 mark for correct reactants and products,
         1 mark for correct balancing]*
You could also write this equation as $C_5H_{10} + 7½O_2 \rightarrow 5CO_2 + 5H_2O$.
9 a) i) The volume of gas had not changed for three days, showing
         that the reaction had finished *[1 mark]*.
      ii) (32.0 − 25.6) ÷ 32.0 = 0.200
         0.200 × 100 = **20.0% oxygen**
         *[2 marks for the correct answer, otherwise 1 mark
         for using the correct method]*
      iii) The overall change in volume would be unaffected
         [1 mark]. However, the time taken for the volume to
         stop changing would decrease *[1 mark]*. A catalyst
         speeds up the reaction rate, but does not affect
         the amount of product formed *[1 mark]*.
   b) i) hydrated iron(III) oxide *[1 mark]*
      ii) **B** *[1 mark]*
      iii) e.g. oiling / greasing *[1 mark]*
      iv) zinc *[1 mark]*
10 a) Using anhydrous copper sulfate would be unsuitable
      [1 mark] because dissolving anhydrous copper sulfate gives
      a temperature increase/is exothermic *[1 mark]*. Any of the
      other three chemicals would be suitable for the chemical
      pack as they all decrease the temperature/are endothermic
      [1 mark]. Ammonium sulfate would be the best choice,
      as it causes the largest temperature decrease *[1 mark]*.
   b) Insulation helps to prevent energy being exchanged with/
      lost to the surroundings *[1 mark]*, which would affect
      the measured temperature change *[1 mark]*.
   c) E.g. she used the same number of moles of each solid
      [1 mark] and the same volume of water *[1 mark]*.
   d) Q = m × c × ΔT
      Q = 50 × 4.2 × (22 − 15)
      Q = 1470 J = **1.5 kJ**
      *[3 marks for the correct answer, otherwise 1 mark for the
      correct equation, 1 mark for the correct value for ΔT]*
   e) M_r(NaCl) = 23 + 35.5 = 58.5
      Q = ΔH × n
      n = mass ÷ M_r(NaCl) = 25 ÷ 58.5 = 0.4273... mol
      Q = +4 × 0.4273...
      Q = 1.709... = **1.7 kJ**
      *[3 marks for correct answer, otherwise 1 mark for correct
      M_r of NaCl, 1 mark for correct moles (n) of NaCl]*
   f) M_r of NH_4NO_3 = (14 × 2) + (1 × 4) + (16 × 3) = 80
      moles of NH_4NO_3 = 5 ÷ 80 = 0.0625 = **0.06 moles**
      *[2 marks for correct answer, otherwise 1 mark for
      correct working]*
   g) Any two from: e.g. it would have a high melting
      point. / It would have a high boiling point. / It would
      conduct electricity when molten/dissolved but not
      when solid *[2 marks — 1 for each sensible answer]*.
These are typical properties of ionic compounds.

Set A — Paper 2

1 a) 4 *[1 mark]*

b) 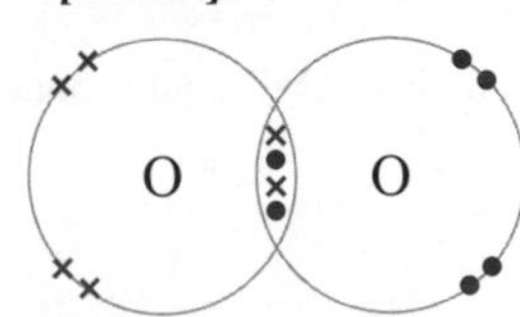

[2 marks — 1 mark for correct number of outer shell electrons on each atom, 1 mark for correct shared electrons]

c) liquid *[1 mark]*

d) Carbon dioxide will have a higher boiling point than oxygen. Carbon dioxide is a larger molecule / has a higher relative molecular mass *[1 mark]*, so the intermolecular forces are stronger *[1 mark]* and so require more energy to break *[1 mark]*.

e) Volume in $dm^3 = 105.6 \div 1000 = 0.1056\ dm^3$
Moles of CO_2 = Volume ÷ Molar Volume
$= 0.1056 \div 24 = 0.0044$ mol
Moles of CO_2 = Moles of $ZnCO_3$
Moles of $ZnCO_3 = 0.0044$ mol
Mass of $ZnCO_3$ = Moles ÷ M_r of $ZnCO_3$
$= 0.0044 \times 125 =$ **0.55 g**
[3 marks for correct answer, otherwise 1 mark for correct method, 1 mark for correct moles of $ZnCO_3$]

2 a) C *[1 mark]*

b) A *[1 mark]*

c) B *[1 mark]*

d) B *[1 mark]*

e) fractional distillation *[1 mark]*

3 a) E.g.

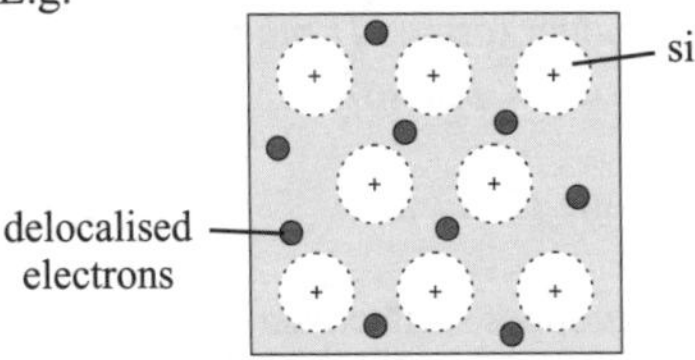

[3 marks — 1 mark for a regular arrangement of metal ions, 1 mark for delocalised electrons, 1 mark for labels]

b) The delocalised electrons are able to move through the metal structure and carry electric charge *[1 mark]*.

c) The layers of ions/atoms within silver are able to slide over each other *[1 mark]*.

4 a) sodium + water → sodium hydroxide + hydrogen *[1 mark]*

b) It is alkaline *[1 mark]*.

c) Potassium is more reactive than sodium *[1 mark]*, because the attraction between the nucleus and the outer electron decreases down Group 1 *[1 mark]*. This means that potassium's outermost electron can be lost more easily *[1 mark]*.

5 a) i) Mass of magnesium = 35.97 g – 35.25 g = 0.72 g
Mass of oxygen = 36.45 g – 35.97 g = 0.48 g
Moles of magnesium = 0.72 g ÷ 24 g/mol = 0.03 mol
Moles of oxygen = 0.48 g ÷ 16 g/mol = 0.03 mol
Simplest whole number ratio:
$(0.03 \div 0.03):(0.03 \div 0.03) = 1:1$
Empirical formula = **MgO**
[1 mark for calculating the mass of Mg, 1 mark for calculating the mass of O, 1 mark for the correct number of moles of magnesium and oxygen, 1 mark for finding the simplest whole number ratio and 1 mark for correct empirical formula]

ii) E.g. not all the magnesium has reacted to form magnesium oxide *[1 mark]*.

b) 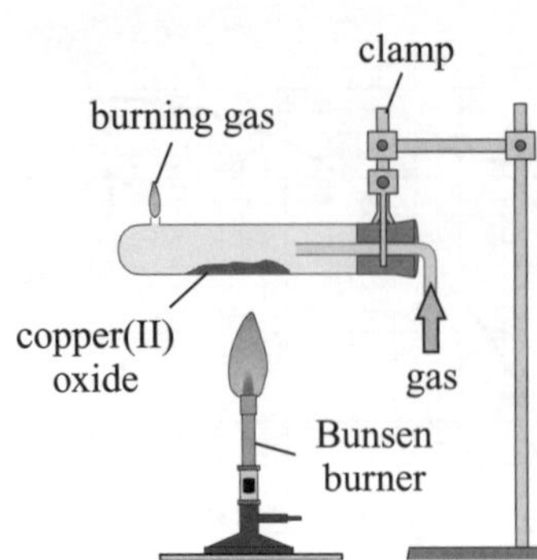

[1 mark for the correct apparatus, 1 mark for labels]

6 a) E.g. measure the potassium hydroxide solution using a pipette and pipette filler rather than with a measuring cylinder *[1 mark]*.

b) E.g. pour the acid below eye level / wear safety goggles *[1 mark]*.

c)

Experiment	Initial reading / cm³	Final reading / cm³	Titre Volume / cm³
1	1.20	18.80	17.60
2	3.35	21.15	17.80
3	1.50	19.20	17.70

[2 marks, 1 mark for each]

d) To make sure that the result was reliable/repeatable / the results were concordant *[1 mark]*.

e) Moles = volume (dm³) × concentration (mol/dm³)
$(17.60 \div 1000) \times 0.0250 =$ **0.000440 moles**
[2 marks for correct answer, otherwise 1 mark for multiplying the volume of acid by its concentration]

f) $0.000440 \times 2 =$ **0.000880 moles** *[1 mark]*
From the equation you can see that one mole of H_2SO_4 reacts with two moles of KOH — so twice the number of moles of KOH will react. If you got the answer to part e) wrong, you can still have the mark here for correctly multiplying your answer by two.

g) Concentration (mol/dm³) = moles ÷ volume (dm³)
$0.000880 \div (50.0 \div 1000) =$ **0.0176 mol/dm³**
[2 marks for correct answer, otherwise 1 mark for dividing the number of moles of potassium hydroxide by its volume]
If you got the answer to part f) wrong, you can still have the marks here for doing the right calculations with your answer.

7 a) i) –OH *[1 mark]*

ii) If the temperature is increased too far above 30 °C, the enzyme in yeast will denature *[1 mark]*. This denaturing would cause the fermentation reaction to stop *[1 mark]*.

iii) combustion *[1 mark]*

iv) carbon dioxide and water *[1 mark]*

b) i)

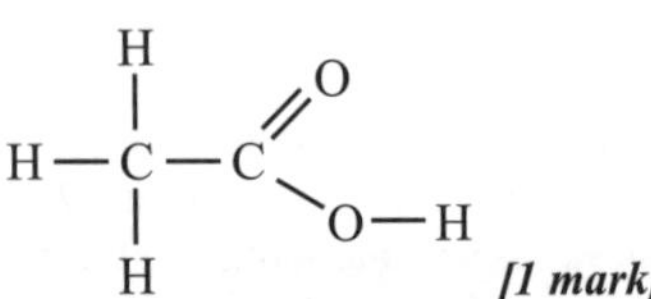

[1 mark]

ii) E.g. by oxidising the ethanol using an oxidising agent such as potassium dichromate(VII) / by microbial oxidation. *[1 mark]*

iii) M_r of $CH_3COOH = (12 \times 2) + (1 \times 4) + (16 \times 2) = 60$
Total relative mass of oxygen = $16 \times 2 = 32$
$(32 \div 60) \times 100 = 53.333...\% =$ **53.3%**
[2 marks for the correct answer, otherwise 1 mark for the correct M_r of ethanoic acid]

8 a) i) Enthalpy change (ΔH) = Total energy absorbed to
        break bonds – Total energy released in making bonds
        Total bond energy in ammonia =
        2 × (3 × 391 kJ/mol) = 2346 kJ/mol
        Total bond energy in hydrogen =
        3 × 436 kJ/mol = 1308 kJ/mol
        –97 = (Total bond energy in nitrogen + 1308) – 2346
        Total bond energy in nitrogen = –97 + 2346 – 1308
        = 941 kJ/mol
        *[4 marks for correct answer, otherwise 1 mark for
        the correct equation/method, 1 mark for the correct total
        bond energy of ammonia, 1 mark for the correct total bond
        energy of hydrogen]*
   ii) The temperature of the surroundings would increase *[1 mark]*
        because the reaction is exothermic /
        transfers energy to the surroundings *[1 mark]*.

   b)   E.g.

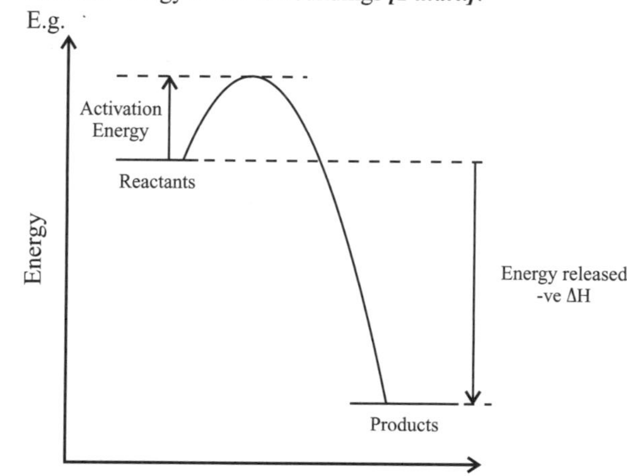

        *[1 mark for the products being lower in energy than
        the reactants, 1 mark for the correct labels]*

9 a)   Calcium nitrate = Soluble
        Lead(II) sulfate = Insoluble
        Sodium hydroxide = Soluble
        Ammonium chloride = Soluble
        [1 mark for any 2 correct, 2 marks for all 4 correct]
   b)   sodium carbonate *[1 mark]*
   Sodium carbonate is soluble in water, so dissolves to become a solute.
   Copper carbonate is insoluble in water, so does not dissolve.
   c)   A solution where the maximum amount of
        solute has been dissolved, so no more solute
        will dissolve in the solution *[1 mark]*.

d) i)

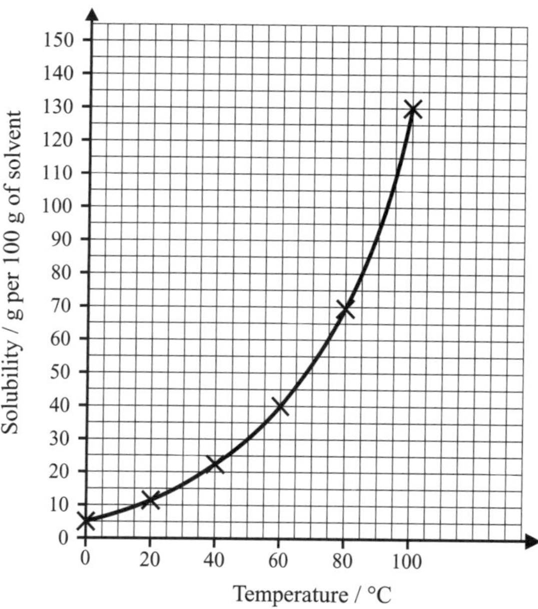

   *[1 mark for plotting the correct points, 1 mark for
   correct curve, 1 mark for correct axes labels]*
   ii) 85 °C *[1 mark]*
   iii) Solubility will not be affected *[1 mark]*.
        The volume of solvent added does not have
        an effect on solubility (solubility is measured
        per a certain volume) *[1 mark]*.

Set B — Paper 1

1 a) A compound formed from hydrogen and carbon *[1 mark]* only *[1 mark]*.

b) fractional distillation *[1 mark]*

c) **B** *[1 mark]*

d) **C** *[1 mark]*

Fractions containing smaller molecules drain out of the fractionating column further up than fractions containing larger molecules.

e) **C** *[1 mark]*

f) i) **D** *[1 mark]*

ii)

Gas	Approximate percentage
Nitrogen	78%
Oxygen	21%
Argon	1%
Carbon dioxide	0.04%

[1 mark for both gases, 1 mark for both percentages]

2 a) i) A nucleus of protons and neutrons *[1 marks]* orbited by electrons *[1 mark]*.

ii)

Particle	Relative mass	Relative charge
Proton	1	+1
Neutron	1	0
Electron	0.0005	−1

[1 mark for both masses, 1 mark for both charges]

b)

Particle	Number in atom
Proton	3
Neutron	4
Electron	3

[1 mark for all 3 answers]

c) i) $^{40}_{20}Ca$ and $^{48}_{20}Ca$ are isotopes because they are atoms of the same element with the same number of protons but differing numbers of neutrons *[1 mark]*.

ii) $(6 \times 7.6) + (7 \times 92.4) = 692.4$

$692.4 \div (7.6 + 92.4) = \textbf{6.9}$

[3 marks for the right answer, otherwise 1 mark multiplying both isotopic masses by their percentage abundance, 1 mark for dividing by the total percentage abundance]

3 a) The components of a mixture are not chemically bonded together *[1 mark]*, but the components/atoms/ions of a compound have chemical bonds between them *[1 mark]*.

b) Propanone and water: Distillation/fractional distillation could be used *[1 mark]*, as propanone and water have different boiling points *[1 mark]*.
Sulfur and water: Filtration could be used *[1 mark]* as sulfur is insoluble in water *[1 mark]*.

c) E.g. Add enough water to the mixture to dissolve the copper sulfate *[1 mark]*. Filter the mixture into a conical flask through a filter paper in a funnel *[1 mark]*. Solid sulfur will be left on the filter paper *[1 mark]*. Pour the solution into an evaporating dish and heat *[1 mark]*. When you see crystals start to form/some of the water has evaporated, remove from the heat *[1 mark]*. Leave the solution to cool and for crystals of copper sulfate to form *[1 mark]*.

4 a) A — sodium chloride
B — methane
C — copper
[2 marks for all three answers correct, otherwise 1 mark for one answer correct]

b) The sodium atoms each lose an electron to form Na^+ ions *[1 mark]*. The chlorine atoms each gain an electron to form Cl^- ions *[1 mark]*.

c) The electrostatic attraction between oppositely charged ions is very strong *[1 mark]*. This means a large amount of energy is required to break them apart / overcome the attraction *[1 mark]*.

d) E.g. substances with giant covalent structures have much higher melting points than substances made of small molecules *[1 mark]*. In order to melt a substance made of small molecules, only weak intermolecular forces need to be broken *[1 mark]*. In order to melt a substance with a giant covalent structure, strong covalent bonds need to be broken *[1 mark]*.

e) i) M_r of oxygen = $16 \times 2 = 32$
Moles of oxygen = mass $\div M_r = 2.88 \div 32 = \textbf{0.09 mol}$
[2 marks for the correct answer, otherwise one mark for using the correct molecular mass for oxygen]

ii) A neutralisation reaction *[1 mark]* will occur, because metal oxides are basic *[1 mark]*.

5 a) i) E.g. the dye in the ink would run up the paper and affect the results. / The ink would dissolve in the solvent *[1 mark]*.

ii) The compounds in the food colouring may not be able to be separated using this particular solvent. / The spot may be produced by a combination of compounds *[1 mark]*.

iii) To stop the solvent from evaporating *[1 mark]*.

iv) E.g.

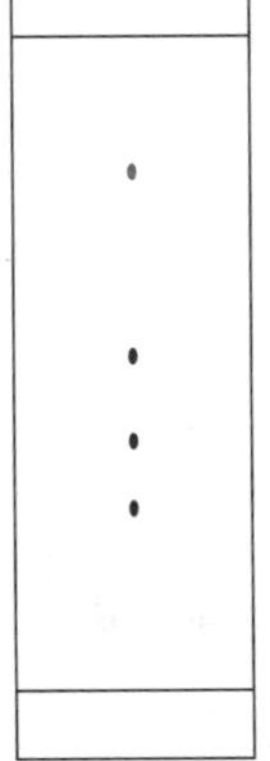

[1 mark for 4 spots on the chromatogram, 1 mark for spots being in a vertical line]

b) i) $R_f = 7.9 \div 9.5 = \textbf{0.83}$
[3 marks for correct answer, otherwise 1 mark for dividing the measurement by 9.5, 1 mark for answer to 2 significant figures]

ii) It is a mixture of the yellow and blue dyes *[1 mark]*. Two spots produced by the green dye have the same R_f values/ are at the same height as spots from the yellow and blue dyes *[1 mark]*.

6 a) E.g. nitrogen oxides are created from a reaction between nitrogen and oxygen in the air *[1 mark]*. This can happen in the internal combustion engines of vehicles *[1 mark]*. As the number of vehicles increases, the emissions of nitrogen oxides also increase *[1 mark]*.

b) E.g. $11.3 + ((11.3 \times 14.0) \div 100) = 12.882 = \textbf{12.9 Mt}$
[1 mark for correct answer, 1 mark for rounding to 3 significant figures]

c) The acidity might increase, as the level of nitrogen oxides increases *[1 mark]*. This is because nitrogen oxides mix with clouds to form nitric acid, which falls as acid rain *[1 mark]*.

d) Reducing the amount of sulfur in fuels may lead to less acid rain being produced *[1 mark]*. This is because sulfur dioxide is released during the combustion of fossil fuels that contain sulfur impurities *[1 mark]*. The sulfur dioxide then mixes with clouds to form dilute sulfuric acid, which falls as acid rain *[1 mark]*.

7 a) i) To increase the rate of the reaction *[1 mark]*.

ii) To ensure that the acid is completely reacted *[1 mark]*.

iii) Solid crystals of magnesium sulfate will form *[1 mark]*.

b) Test: e.g. test the pH of the solution using pH indicator. / Test the pH of the solution using litmus paper *[1 mark]*. Result: e.g. the pH would be 7/solution would be neutral. / The paper would turn green *[1 mark]*.

c) H^+ ion *[1 mark]*

d) Oxidised species: magnesium/Mg *[1 mark]*
Reason: Magnesium/Mg has lost electrons *[1 mark]*.

e) i) More bubbles will be formed in the tube containing magnesium *[1 mark]*. Magnesium is more reactive than zinc and therefore reacts more vigorously with the acid *[1 mark]*.

ii) Moles of Zn = mass ÷ A_r = 1.25 ÷ 65 = 0.01923... mol
Molar ratio of Zn : $ZnSO_4$ = 1:1, so
Moles of $ZnSO_4$ = moles of Zn = 0.01923... mol
Theoretical yield of $ZnSO_4$ = moles × M_r = 0.01923... × 161
$\qquad$ = 3.096...
Percentage yield = (actual yield ÷ theoretical yield) × 100
$\qquad$ = (1.03 ÷ 3.096...) × 100
$\qquad$ = 33.26...
$\qquad$ = **33.3%**
[4 marks for the correct answer, otherwise 1 mark for the correct number of moles of Zn and $ZnSO_4$, 1 mark for finding the theoretical yield of $ZnSO_4$ and 1 mark for the equation for percentage yield]

8 a) 15 seconds *[1 mark]*
The temperature reading at 15 seconds is anomalous because it doesn't fit in with the trend in the rest of the data.

b) Energy is taken in from the surroundings *[1 mark]*.
There is a decrease in temperature *[1 mark]*.

c) E.g. Add a set amount (e.g. 25 cm³) of a known concentration of citric acid to a beaker and add a set amount (e.g. 25 cm³) of a known concentration of sodium hydrogen carbonate to a different beaker *[1 mark]*. Place the beakers in a water bath set to a known temperature (e.g. 25 °C) and leave the beakers in the water bath until they are both at the same temperature *[1 mark]*. Add the citric acid and the sodium hydrogen carbonate to an insulated polystyrene cup with a lid *[1 mark]*. Measure the temperature of the mixture using a thermometer, and record the lowest temperature *[1 mark]*. Repeat the experiment two more times, keeping all the variables the same to make sure that the result is reliable *[1 mark]*. Repeat the experiment with a different concentration of citric acid, keeping everything else the same *[1 mark]*.

9 a) potassium / K^+ *[1 mark]*

b) Add a few drops of sodium hydroxide solution to the solutions *[1 mark]*. The Fe^{2+} ions in Solution 2 will form a (sludgy) green precipitate *[1 mark]* and the Fe^{3+} ions in Solution 3 will form a (reddish) brown precipitate *[1 mark]*.

c) i) To remove other ions that may cause a positive result (e.g. sulfite/carbonate ions) *[1 mark]*.

ii) The flame test produced a yellow flame, which means that sodium ions must be present *[1 mark]*. The solution remained clear during the sulfate ion test, which means that no sulfate ions are present *[1 mark]*. The halide test gave a cream precipitate, which means bromide ions must be present *[1 mark]*. The compound in Solution 4 must therefore be sodium bromide *[1 mark]*.

10 a) The solution would turn orange *[1 mark]* as the bromine/bromide in KBr is displaced by chlorine *[1 mark]*.

b) E.g. moles of Rb = mass ÷ M_r = 3.40 ÷ 85 = 0.04 mol
The compound will have the formula RbX
Equation = $2Rb + X_2 \rightarrow 2RbX$
There are 2 molecules of Rb and 2 molecules of RbX in the equation, therefore they are in a 1:1 ratio.
This means there must be 0.04 mol of RbX.
M_r(RbX) = mass ÷ moles = 4.82 ÷ 0.04 = 120.5
A_r(Rb) = 85
So A_r(X) = 120.5 − 85 = 35.5
The unknown halide is **chlorine**.
[4 marks for the correct answer, otherwise 1 mark for the correct number of moles, 1 mark for the formula RbX and 1 mark for the correct M_r of the product]

c) i) $CuBr_2$ *[1 mark]*

ii) M_r of butene = (12 × 4) + (1 × 8) = 56
M_r of bromine = 80 × 2 = 160
Moles of butene = 3.36 ÷ 56 = 0.06
Moles of bromine = 3.36 ÷ 160 = 0.021
The equation shows that they react in the ratio 1:1, so reactant in excess = butene
[1 mark for comparing the molecular masses of butene and bromine, 1 mark for comparing the number of moles of each, 1 mark for identifying the reactant in excess]

d) i) This reaction must occur in the presence of ultraviolet light *[1 mark]*.

ii)

[1 mark for the correct structure, 1 mark for the correct number of electrons]

iii) 1-chloropropane *[1 mark]* and 2-chloropropane *[1 mark]*

Set B — Paper 2

1 a)   **C** *[1 mark]*
  b)   **D** *[1 mark]*
  c)   **A** *[1 mark]*
  d)   **B** *[1 mark]*
  e)   E.g. A decrease in temperature. This would shift the equilibrium in the exothermic direction, increasing the yield of sulfur trioxide. / An increase in pressure. This would encourage the reaction which produces fewer moles of gas.
    [1 mark for change in conditions, 1 mark for justification]

2 a)   Any two from: e.g. wear safety goggles / use a heat proof mat under the apparatus / wear a lab coat *[2 marks]*
  b)   Molar ratio of $CuO : CuCO_3 = 1:1$, so
    moles of $CuCO_3$ = moles of CuO = 0.050 mol
    M_r of $CuCO_3 = 63.5 + 12 + (16 \times 3) = 123.5$
    Mass of $CuCO_3$ = moles $\times M_r = 0.050 \times 123.5$
    $= 6.175 \approx$ **6.2 g**
    [2 marks for the correct answer, otherwise 1 mark for the correct M_r of $CuCO_3$]
  c) i)   reduction *[1 mark]*
    ii)   It is higher in the reactivity series/more reactive than copper *[1 mark]*.

3 a) i)   Both air and water are necessary *[1 mark]*.
    ii)   The nail attached to zinc is likely to have no rust as zinc is more reactive than iron (and so will be oxidised instead) *[1 mark]*.
  b) i)   Steels are iron alloys because they contain atoms of different elements that have been added to the iron *[1 mark]*.
    ii)   They are often too soft/malleable *[1 mark]*.
    iii)   Use: e.g. taps/sinks/kitchen equipment/food storage containers/medical equipment *[1 mark]*.
      Explanation: e.g. stainless steel won't corrode when the object is regularly exposed to both water and air *[1 mark]*.

4 a) i)   lead sulfate *[1 mark]*
    ii)   To ensure that there are no other ions that could react *[1 mark]*.
    iii)   E.g. an oven / a desiccator *[1 mark]*
  b) i) $Ca(NO_3)_2$ *[1 mark]*
    ii)   Volume of nitric acid in $dm^3 = 450\ cm^3 \div 1000 = 0.450\ dm^3$
      Moles of nitric acid $= 0.450\ dm^3 \times 1.5\ mol/dm^3 =$ **0.675 mol**
      [2 marks for the correct answer, otherwise 1 mark for converting cm^3 to dm^3]
  c)   Mass of $Zn(NO_3)_2.xH_2O$
    = (starting mass of crucible + sample) – mass of empty crucible
    $= 65.233 - 61.500 = 3.733$ g
    Mass of anhydrous $Zn(NO_3)_2$
    = (final mass of crucible + sample) – mass of empty crucible
    $= 63.876 - 61.500 = 2.376$ g
    Mass of water lost
    = mass of $Zn(NO_3)_2.xH_2O$ – mass of anhydrous $Zn(NO_3)_2$
    $= 3.733 - 2.376 = 1.357$ g
    Moles of water lost = mass $\div M_r = 1.357 \div 18$
    $= 0.07538...$ mol
    Moles of anhydrous $Zn(NO_3)_2$ = mass $\div M_r = 2.376 \div 189$
    $= 0.01257...$ mol
    Molar ratio of salt : water $= 0.01257... : 0.07538...$
    $= 1 : (0.07538... \div 0.01257...)$
    $= 1 : 6$
    x = 6
    [4 marks for the correct answer, otherwise 1 mark for the correct masses of anhydrous $Zn(NO_3)_2$ and water lost, 1 mark for the moles of $Zn(NO_3)_2$ and 1 mark for the moles of water lost]

5 a)   An alkali dissolves in water and releases OH^- ions/gives a solution with a pH greater than 7 *[1 mark]*.
  b) $10\ cm^3 = 0.01\ dm^3$
    Moles of KOH $= 0.01 \times 3 = 0.03$ moles
    Final volume needed $= 0.03 \div 1.2 = 0.025\ dm^3 = 25\ cm^3$
    Volume of water to add $= 25 - 10 =$ **15 cm³**
    [3 marks for correct answer, otherwise 1 mark for correct moles of KOH, 1 mark for calculating that the total volume needs to be 25 cm^3]
  c)   Measure out a known volume of the potassium hydroxide solution using a pipette and put it in a conical flask *[1 mark]*. Add a few drops of a suitable indicator (e.g. phenolphthalein or methyl orange) to the potassium hydroxide *[1 mark]*. Use a burette to slowly add the hydrochloric acid to the potassium hydroxide, swirling the conical flask regularly, using a white tile to make sure the colour change is easy to see *[1 mark]*. Stop the titration when the indicator changes colour, recording the volume of hydrochloric acid required to cause this colour change — this is when the potassium hydroxide is completely neutralised *[1 mark]*. Repeat the titration with the same volume of potassium hydroxide solution, but add the acid one drop at a time close to the end point, recording the exact volume of hydrochloric acid required to neutralise the potassium hydroxide *[1 mark]*. Repeat the titration several times until you obtain concordant results *[1 mark]*.
  d)   Moles of KOH used $= 1.2\ mol/dm^3 \times 0.025\ dm^3 = 0.03$ mol
    The ratio between HCl and KOH in the reaction equation is 1:1, so 0.03 moles of KOH must react with 0.03 moles of HCl.
    Mean titre of HCl $= (12.80 + 12.75 + 12.85) \div 3 = 12.80\ cm^3$
    Concentration of HCl $= 0.03 \div (12.80 \div 1000) =$
    2.34 mol/dm³
    [4 marks for correct answer, otherwise 1 mark for correct moles of KOH, 1 mark for correct moles of HCl, 1 mark for calculating the mean titre of HCl]

6 a) i)   ethanoic acid *[1 mark]*
    ii)

$$\begin{array}{ccccccc}
 & H & O & & H & H & H & H \\
 & | & \| & & | & | & | & | \\
H\!-\!\!&C\!\!&-\!\!C\!\!&-O-\!\!&C\!\!&-\!\!C\!\!&-\!\!C\!\!&-\!\!C\!\!-H \\
 & | & & & | & | & | & | \\
 & H & & & H & H & H & H
\end{array}$$

    [1 mark]

This is the ester formed from butan-1-ol and ethanoic acid, but if you drew the ester formed from butan-2-ol and ethanoic acid, then you'd still get the mark.

  b) i)   Method: bubble the gas through limewater *[1 mark]*. Expected result if carbon dioxide is present: the limewater will turn cloudy *[1 mark]*.

Limewater is the name given to a solution of calcium hydroxide.

    ii)   They could be difficult to count accurately/the student might miss some of the bubbles *[1 mark]*.
    iii)   E.g. use solutions with lower concentrations *[1 mark]*. Fewer particles in the same volume of solution *[1 mark]* means that the frequency of collisions decreases *[1 mark]*. / Use a lower temperature *[1 mark]*. This makes particles move more slowly *[1 mark]* and collision frequency and energy of collisions will decrease *[1 mark]*.
    iv)   E.g. she could get someone else to repeat the experiment to see if they obtain the same results *[1 mark]*.

v) E.g.

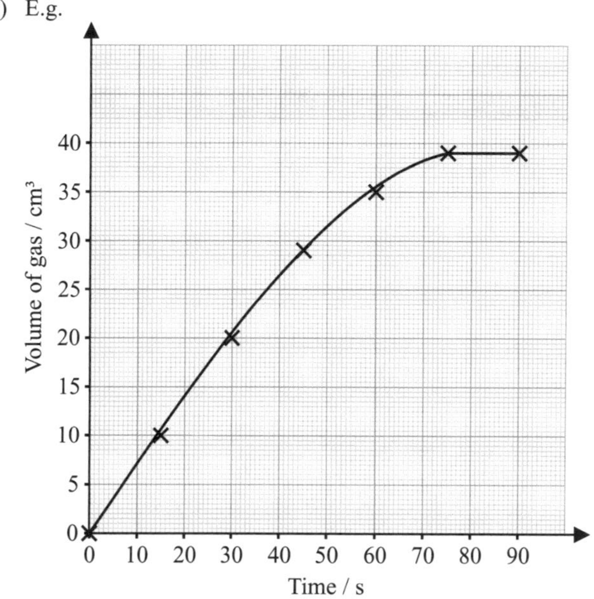

[1 mark for sensible scale using over half of grid, 1 mark for correct axis titles, 1 mark for all points plotted correctly, 1 mark for line of best fit]

vi)

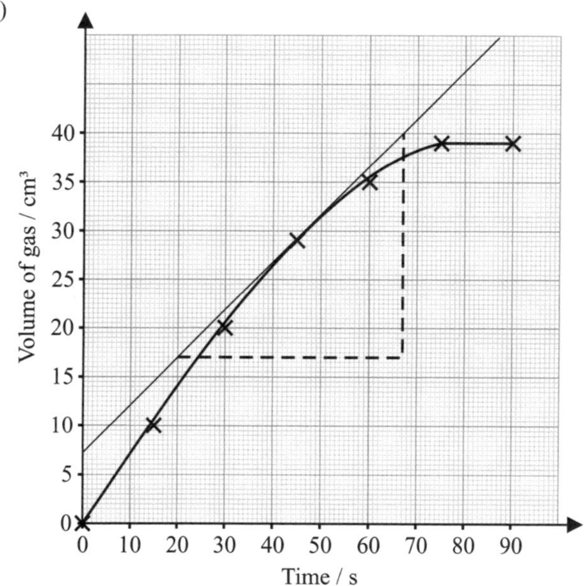

Change in volume = 40 − 17 = 23 cm³
Change in time = 67 − 20 = 47 s
Gradient = 23 ÷ 47 = 0.4893... = 0.49
Rate of reaction = 0.49 cm³/s

[1 mark for tangent drawn correctly, 1 mark for correct method of calculating gradient, 1 mark for gradient in the range 0.45-0.55]

Don't worry if your triangle isn't in exactly the same place or if you have used slightly different numbers. As long as your tangent was sensible, your final answer should be within the range of acceptable answers.

7 a) In electrolysis, the ions in the substance need to be free to move *[1 mark]*.

b) Electrode: anode/positive electrode *[1 mark]*
Explanation: Oxidation involves the loss of electrons *[1 mark]*. The positively charged anode attracts negative ions and removes electrons from them *[1 mark]*.

c) E.g. potassium hydroxide is cheaper/easier to melt as its melting point is lower than that of potassium carbonate *[1 mark]*.

d) i) It is inert *[1 mark]*.

ii) $2H^+_{(aq)} + 2e^- \rightarrow H_{2(g)}$
[2 marks — 1 mark for correct reactants and products, 1 mark for correct balancing and state symbols]

The salt being electrolysed must be potassium carbonate or potassium hydroxide. Potassium is higher in the reactivity series than hydrogen, so hydrogen is produced at the cathode.

iii) Moles of oxygen = 4.8 g ÷ 32 g/mol = 0.15 mol
Volume of oxygen = 0.15 mol × 24 = 3.6 dm³ = **3600 cm³**
[2 marks for the correct answer, otherwise 1 mark for the correct number of moles of oxygen]

iv) They could hold a glowing splint over the reaction vessel *[1 mark]*. If the gas is oxygen, the splint will relight *[1 mark]*.